BEFORE DARWIN

BEFORE DARWIN

Reconciling God and Nature

Keith Thomson

Yale University Press *New Haven and London*

Published with assistance from the Annie Burr Lewis
Fund. First published in the United States in 2005 by Yale
University Press. First published in Great Britain in 2005
by HarperCollins Publishers.

Set in Sabon by Rowland Phototypesetting Ltd., Bury St.
Edmunds, Suffolk. Printed in the United States of
America.

Library of Congress Control Number: 2004117193
ISBN 0-300-10793-5 (cloth : alk. paper)

A catalogue record for this book is available from the
British Library.

The paper in this book meets the guidelines for perma-
nence and durability of the Committee on Production
Guidelines for Book Longevity of the Council on Library
Resources.

10   9   8   7   6   5   4   3   2   1

In memory of
The Reverend Ronald William Thomson (1908–2002)

'Knowledge ... concerning God ... may be obtained by the contemplation of his creatures; which knowledge may be truly divine in respect of the object, and natural in respect of the light.'

Francis Bacon, *The Advancement of Learning*, 1605

'All religious systems ... are subject to great and insuperable difficulties. Each disputant triumphs in his turn; while he carries on an offensive war, and exposes the absurdities, barbarities, and pernicious tenets of his antagonists. But all of them, on the whole, prepare a complete triumph for the sceptic; who tells them, that no system ought ever to be embraced with regard to such subjects: for this plain reason, that no absurdity ought ever to be assented to with regard to any subject.'

David Hume,
Dialogues Concerning Natural Religion, 1779

'I do not think I hardly ever admired a book more than Paley's "Natural Theology". I could almost formerly have said it by heart.'

Charles Darwin,
letter to John Lubbock, 15 November, 1859

CONTENTS

PREFACE   xi

1 Charles Darwin and William Paley   1

2 An Age of Science, An Age of Reason   21

3 Problems at Home   45

4 John Ray: Founding Father   59

5 Difficulties with the Theory, and the
   Argument Extended   83

6 Fossils and Time: Dr Plot's Dilemma   111

7 Sacred Theories   138

8 Unfinished Business: Mountains
   and the Flood   174

9 This is Atheism   197

10 Gosse's Dilemma and Adam's Navel   223

11 Good and Evil: Concerning the Mind
   of God   232

12 Paley, Malthus and Darwin   245

13 The Beginning of the End   266

APPENDIX:
   The Account of Creation in Genesis   281

ACKNOWLEDGEMENTS   283

BIBLIOGRAPHY   285

NOTES   295

INDEX   305

PREFACE

When Charles Robert Darwin entered the University of Cambridge in 1828, it was not with the expectation of studying science but rather with every intention of becoming a Church of England priest. He studied theology, philosophy, classics and a little mathematics. We do not know with what enthusiasm he faced the prospect of actually being ordained; his early writings show a conventional set of beliefs and a noticeable lack of religious zeal. Over the following decades, however, everything changed for Darwin. He found that his science, his intellect and his faith had created a set of personal conflicts that he could not resolve.

Charles Darwin's dilemma was the inevitable consequence of pursuing a system of rational enquiry like science that seeks to explain all apparent mysteries in terms of lawful, natural phenomena. He was forced to choose an intellectual path between a belief that the world was created by God exactly as it is now, and probably some 6,000 years ago, and the growing evidence that the earth and life upon it have evolved slowly, without supernatural assistance, for some 4.5 billion years. In fact, in various guises and under different names, what we call 'evolution' had been in the air for at least two hundred years before the publication in 1859 of his *On the Origin of Species by Means of Natural Selection and the Preservation of favoured Races in the Struggle for Life*. Since about 1650, the development of a whole range of new sciences had posed threat after threat to the mysteries of revealed religion. 'Evolution' is part of a much broader and older enquiry and a deeper contest for our intellectual commitment, a contest between a world system that expects every part of the cosmos ultimately to be explainable in terms of natural properties and processes and one that maintains the existence of a fundamental core of unknowability, of supernatural mystery and the controlling hand of an eternal non-worldly Being. This may be humankind's oldest intellectual puzzle, and this book will attempt to trace one small part of it.

It has not always been necessary to choose between one side or the other: science or religion, reason or mystery. In the age before Darwin, many powerful clerics were also notable scientific scholars and leading scientists were often at least conventionally pious. For them, science and religion could share a common philosophical basis with the premise that a careful, rational study of nature, instead of denying God, would confirm that all life is, after all, the product of God's unique creation. *Natural Theology* and its counterpart in the geological context, *Physico-Theology*, provided an intellectual framework that both embraced science and kept it at bay. Indeed, natural theologians believed that a study of God's handiwork constituted a proof of the very existence of God. Believers who were scientists welcomed natural theology because it gave their endeavours a framework within which to operate. Deist and Christian alike could find much to favour in a movement that sought to discover God through rational study without depending on a belief in miracles or insisting on the literal truth of the Bible. Some theologians naturally worried that this new movement would risk flirting too seductively with material explanations of the world and preferred to remain with the relative safety of the authority of the Bible and revelation. Nonetheless, in one form or another, natural theology has maintained a currency to the present day. Its last and greatest expression was in the classic work of 1802 by Reverend William Paley, usually known by the short version of its title, *Natural Theology*. Paley's arguments have never really been improved upon. His book affords a starting point from which to trace a story that reaches from the ancient Greeks to Descartes and to the intellectual environment of the seventeenth and eighteenth century from which Darwin sprang.

Paley and his predecessors set themselves a daunting task. As the explanatory power of science had reached further and further into the previously secret places of nature and as it progressed beyond simple description of phenomena to explain the earth and life on it in terms of underlying processes, their arguments had to change in response. Some of what they wrote – what they theorised, even what they thought was fact – might almost seem absurd to us now, but the work of people such as the Reverend Thomas Burnet, the Reverend John Ray, Dr Robert Plot, Dr William Whiston, Dr John Woodward, Dr James Hutton, Dr Erasmus Darwin and countless more, writing between 1665 and 1800, gives us an extraordinary glimpse into minds at the forefront of an epic enquiry. Many were deeply religious; all were consumed by the need to solve this potentially all-consuming challenge to their world view. They were often under immense pressure to conform with orthodoxy, even when their intellects pointed them in other directions. As well as admiring the force and elegance of the writing and the inventiveness of the arguments, we may also variously envy these earlier scholars their certitude, their daring, and often their humility and caution. It also becomes clear that this is not just a struggle between two sides; it is also a series of debates raging *within* both what we now call 'science' and 'religion'.

One of the many ironies about the conflict over the subject of evolution is that as a student Charles Darwin read and admired Paley, including his great work on natural theology. The final theme to this book is to trace out the powerful, direct connection from Paley's belief that natural theology proved the impossibility of 'evolution' to Darwin's championing of a critically important mechanism for the origin of new species.

CHAPTER ONE

Charles Darwin and William Paley

'The Bible tells us how to go to Heaven, but not how the heavens go.'

Galileo Galilei, letter to the Grand Duchess Christina, 1615 (quoting Cardinal Caesar Baronius)

'As it more recommends the Skill of an Engineer to contrive an elaborate Engine so as that there should need nothing to reach his ends in it but the contrivance of parts devoid of understanding . . . so it more sets off the Wisdom of God . . . that he can make so vast a machine [the universe] perform all these many things.'

Robert Boyle, *Free Enquiry into the Nature of Things*, 1688

One enters Christ's College, Cambridge, through a richly carved sixteenth-century gateway and under a pair of painted heraldic beasts, all contrasting markedly with the sober courtyard of grey buildings within. Across the immaculate lawn, on the right-hand side of First Court, is the doorway to Staircase G with, on the first floor, the pair of rooms occupied by the shy young Charles Darwin when, between 1828 and 1831, he studied to become a Church of England cleric.

Much of Darwin's early life, his ambitions and the sources of his inspiration, remain a mystery. He had originally started to train for medicine at Edinburgh but neither the subject nor the intellectual climate of the city suited him and in 1828 he

1

entered Cambridge to prepare for a life as a clergyman instead. With his driving passion for natural history, he may have had in mind a career as a country parson-naturalist in the long tradition that had produced such luminaries as the Reverend John Ray (known as 'the father of natural history') in the seventeenth century and the Reverend Gilbert White, revered chronicler of English country life in the eighteenth. (Darwin's cousin William Darwin Fox was also at Cambridge planning just such a career and soon achieved it, although his scholarly contributions from his quiet country parish were minor.) He may even have aspired to become a university don like his teachers in Edinburgh (Robert Jameson and Robert Grant) or his eventual Cambridge mentor the Reverend Professor John Stevens Henslow, a cleric, a brilliant teacher and a leading botanist and geologist. But if he were to take the route of training for the clergy, there was first the issue of faith.

Darwin had been brought up in the Midlands Unitarianism of his mother (the daughter of Josiah Wedgwood) while his father had long since taken the fashionable road to the Church of England. There is no question of Darwin having had a special 'calling' to be a clergyman. Indeed, when his father insisted that if he would not continue with medicine he must enter the Church, the eighteen-year-old had privately questioned whether he was sufficient of a believer honestly to start down that path, let alone to give witness to his belief in the pulpit.[1] But he needed a respectable profession. Therefore, in the summer of 1827, in his calm, preternaturally rational way, he set out on a research programme to discover whether he could go through with it. Darwin carefully studied the Reverend John Pearson's *Exposition of the Creed* (1659)[2] with one question in mind, and 'as I did not in the least way doubt the strict and literal truth of any word in the Bible, I soon persuaded myself that our Creed must be fully accepted'. Satisfied at a minimal level

Charles Darwin

that he was not being personally or intellectually dishonest, Darwin entered Cambridge.

Darwin failed to complete his clerical training, just as he earlier failed to complete his medical studies. In 1831, armed with a passing degree and financially secure from his mother's estate, he went off for five years' adventuring and discovery on HMS *Beagle*. His religious beliefs then were still quite conventional: 'I remember being heartily laughed at by several of the officers (although themselves orthodox) by quoting the Bible as an unanswerable authority on some point of morality.' By the time he returned his career had taken a different direction, one in which the ceaseless questioning of science gradually replaced the sureties of revealed religion. But from very early on Darwin thought seriously about the developing conflicts between science and religion. As a student preparing to take holy orders, he

3

knew of the challenges posed by early theories of evolution – particularly since one of them was the brainchild of his own grandfather, Erasmus Darwin. Later he would be only too aware that his own theory of natural selection, which he began to formulate as early as 1838, would inevitably contribute to the growing crisis caused by scholars who discovered, behind the apparent miracles of nature, the operation of scientifically definable laws and processes. And the situation was the more personal after his marriage in 1839 to his cousin Emma Wedgwood, who for her whole life was a staunchly believing Christian. At the time of writing *On the Origin of Species*, Darwin had lost his faith as a Christian and thought himself a deist; he died an agnostic and, while we cannot be sure exactly when Darwin first faced the challenges presented to conventional faith by contemporary science, we know that he was well aware of the issues in 1831, because we know what books he read.

In the spring of 1831, the tall, shy, aspiring cleric, a paradoxical mixture of bookish intellectual and outdoorsman with a passion for field sports, found himself with some time on his hands. Although he had achieved a respectable tenth place among students not competing for honours, he would not technically be eligible to graduate. He had to complete his required period of residency before entering the final year of study that would complete his formal preparation for ordination and a career in the Church. There was no question of his joining the 'fast set' at Cambridge and wasting his time with gambling and women. Typically, his mentor the Reverend Henslow prescribed a programme of reading: as a prospective ordinand in the Church of England, the young man of course read theology; as a keen naturalist and collector, especially of beetles, he read in travel and natural science. In his autobiography, written some fifty years later, Darwin singled out books from this period that had been most influential on his intellectual development. These included John Herschel's *Preliminary Discourse on the Study of Natural Philosophy*,[3] a book on the scientific method and the nature of scientific 'proof', and Alexander von Humboldt's *Personal Narrative*[4] of scientific exploration in South America. He also read the Reverend William Paley's *Natural Theology*[5], a treatise on the use of science to prove the existence, and demonstrate the attributes, of God. These three books, although very different from each other in subject matter, each dealt in their own way with the logic, philosophy and methodology of discovery and proof. Humboldt, whose work on the variation of climate with altitude Darwin had read at Edinburgh, helped fire his passion for exploration and discovery, and showed how the natural world could be explained in terms of natural laws. Herschel outlined the essential elements of a rigorous Baconian scientific explanation for any phenomenon, and it has been argued that the structure of Darwin's own *On the Origin of Species*, published in 1859, was composed following Herschel's rules exactly. In *Natural Theology* William Paley applied rigorous logic and a broad knowledge of philosophy to a wide range of contemporary scientific data in order to attempt nothing less than a final proof of the nature of God. This was a work intended to bridge two worlds that had long been threatening to pull apart. It would resolve the conflict that we find still unresolved today between, on the one hand, the world of scientific explanation expressed in definable, measurable, physical properties and natural laws and, on the other, belief in a God who transcends the material world.

William Paley's theological works were well known to all students at Cambridge, where the syllabus included formal study of two of his books. Darwin, who had a particular appreciation for finely argued logic and reason, was examined on Paley in 1830 and said,

I am convinced that I could have written out the whole of the *Evidences* [Paley's *A View of the Evidences of Christianity*, 1794] with perfect correctness . . . the logic of this book and as I may add of his *Natural Theology* gave me as much delight as did Euclid. The careful study of these works, without attempting to learn any part by rote, was the only part of the Academical Course which, as I then felt and as I still believe, was the most use to me in the education of my mind.[6]

Paley's last book, *Natural Theology*, was not a set book for Cambridge examinations.[7] It aimed for a broader audience than theologians alone and has come to occupy a special place in the history of science and religion. The basic premise of the larger movement of the same name was that the glories and complexities of living nature were to be seen as prima facie evidence of the power of God's creative hand. From this viewpoint, which owes its origins among others to the Five Ways of St Thomas Aquinas, there could be no more pious endeavour than to study nature. All the patterns, symmetries and laws of nature were simply the reflection of God's mind. Therefore to study nature was to approach closer to God. Indeed, the deepest study of nature would provide confirmation of God's very existence. Natural science and theology were not at odds, therefore, but complementary. In particular, any kind of evolutionary theory of the kind that had been growing for the previous hundred years – in which the study of nature pointed to different, material causes of life in all its magnificent diversity than the hand of God – would be negated. At the time the young Darwin studied for the Church at Cambridge, as for a hundred years before, natural theology offered a rationale for the reconciliation of what might have seemed to be opposed: the diverse worlds of science and of religion.

This argument has a strong following today among those who

would oppose, or are agnostic about the theory of evolutionary change. But, curiously, a direct connection can be traced between Paley's arguments *against* any kind of evolutionary theories (of which there were many, termed 'transmutation', or 'development' theories, long before Charles Darwin was even born) and the origins of modern scientific thinking *in favour of* evolutionary theory. Darwin's reading of *Natural Theology* in 1831 therefore has a particular resonance for anyone today who is interested in the question of how the apparently separate subjects of science and religion can be made one and, indeed, anyone interested in the historical precedents and intellectual origins of modern evolution.

Although many of the driving intellects of the age were continental – and this particular version of the battle between science and religion was being fought out elsewhere than Britain – this is a story about a peculiarly English part of the phenomenon, set squarely within a long English tradition. Its Englishness was due in large part to the long tradition of the English cleric-naturalists whose science was based in empiricism: from their rural parishes they observed nature and tried to read the word of God in it. They belonged straightforwardly to the new critical age and applied its rules and procedures to their thinking in God's service. With many Church of England livings conveniently tied to the universities of Cambridge and Oxford, they had security and access to nature on the one hand, and on the other were a direct arm of the intellectual work of learning and teaching. Under the influence of the Enlightenment, they preached, they taught, they observed, and they considered it all. William Paley's summation of the arguments of natural theology comes at the end of a great period of learning and adventure, with its freedom to entertain heretical ideas and no little reactionary conservatism, in which the evidence of nature as

7

revealed by science was used to argue for the existence and nature of God.

Naturally, not everyone in the mainstream Church or its many dissenting offshoots approved of natural theology and attempts to prove God through science. The traditional route to discovery of God was through the authority of the Bible, divine revelation and the life of Christ. It was built upon the constancy of faith rather than the shifting ground of science. The paradox and the strength of faith is that it is not susceptible to cold-eyed analysis. No one knows if his faith is the same as another's; almost by definition it cannot be. At the heart of Judeo-Christian tradition, the Bible reveals to us – individually or through the exegesis of its spiritual leaders – all that we need to know about God. It *tells* us that God is the Creator, all-wise and all-good, and is full of internal proofs, one of the greatest being that written in Isaiah 7:10–14: 'Moreover the Lord spake unto Ahaz, saying, Ask thee a sign of the Lord thy God; ask it either in the depth or in the height above. But Ahaz said, I will not ask, neither will I tempt the Lord.' Then God promised the ultimate evidence: 'The Lord himself shall give you a sign; Behold a virgin shall conceive, and bear a son, and shall call his name Immanuel.' For evidence, there were the miracles, both the biblical miracles – especially those wrought by Jesus, healing the sick, raising the dead, as he prophesied in the name of God – and those performed by God through his saints on earth. And the most dramatic demonstration of God's existence and power would have to be the resurrection of Jesus, who in turn gave another real proof when he allowed the doubting Thomas, after the resurrection, to fit his hand into the spear wound in his side. No other authority than these was needed and an unguarded or naive person attempting to find God through the objective evidences of science might risk challenging traditional modes of authority and even be seduced by material explanations of phenomena to an opposite, atheistical position.

One can imagine circumstances when an attempt to prove God's existence would have been heretical. A group of scholars probing nature to see if they can winkle out secrets about God that have been hidden – if God exists, then deliberately hidden – for thousands of years, seems a dangerous idea. In the year 1200, for example, not only would it have guaranteed the Pope's or a bishop's punishment; any sensible person would fear that God himself might exercise a little discipline too; after all, the Bible says 'Ye shall not tempt [test or prove] the Lord thy God'.[8] Or, as St Luke wrote: 'This is an evil generation: they seek a sign.'[9] In any case, to frame a series of questions and statements about his existence, unless done carefully, would be to risk limiting him to the small compass that our understanding allows. It also risked provoking further challenges to the literal truth of the Bible, already a problem in Enlightenment times. Close study of the Bible showed some worrying inconsistencies – two different versions of the Flood, for example, and two of the creation of Woman, together with much that is in disagreement with the facts of modern science. And why was so dramatic a sign as Thomas's encounter with the risen Jesus (John 20: 24–28) not mentioned in any of the other three gospels? There was great danger in holding the Bible up to the same scrutiny, the rigid tests of independence, as used in the accepted methods of science.

But Paley was a man living in an age of science and reason. What might in medieval times have been considered dangerous or blasphemous – to prove something that required no proof – was now both acceptable and necessary. Philosophers such as Locke and Hume had long since exposed the vulnerability of religion founded not in fact but belief or faith, growing out of intuition, inspiration, hopes, fears and even myths. And, when Paley wrote in the dedicatory preface to *Natural Theology* that

William Paley

more and different proofs were needed because of the 'scepticism' concerning the existence and attributes of God 'with which the present times are charged', he could easily have been writing of our own day. The strength of his natural theology was that it did not depend on divine revelations and demanded no leaps of faith and suspensions of disbelief. Miracles are not mentioned in Paley's book. Interestingly enough, neither is the Bible. Instead its argument depends upon a single, central argument in the form of a syllogism, summarised in his beautifully simple first paragraph. One can see instantly that someone like Darwin, who was attracted to logic and eager to see his inner doubts allayed, would be intrigued by this, the opening paragraph of Paley's book:

In crossing a heath, suppose I pitched my foot against a stone, and were asked how the stone came to be there, I might possibly answer that, for any thing I knew to the contrary, it had lain there for ever; nor would it perhaps be very easy to shew the absurdity of this answer. But suppose I had found a watch upon the ground, and it should be enquired how the watch happened to be in that place . . . why should not this answer serve for the watch, as well as for the stone? . . . For this reason, and no other, viz. that, when we come to inspect the watch, we perceive, (what we could not discover in the stone,) that its several parts are framed and put together for a purpose, e.g. that they are so formed and adjusted as to produce motion, and that motion so regulated as to point out the hour of the day; that, if the several parts had been differently shaped from what they are, of a different size from what they are, or placed after any other manner, or in any other order . . . none which would have answered the use, that is now served by it . . . This mechanism being observed, (it requires indeed an examination of the instrument, and perhaps some previous knowledge of the subject, to perceive and understand it) . . . the inference, we think, is inevitable; that the watch must have had a maker; that there must have existed, at some time and at some place or other, an artificer or artificers who formed it for the purposes, which we find it actually to answer; who comprehended its construction, and designed its use.

In this analogy the watch stands for all living organisms; the 'who', of course, is God. A modern wristwatch is a dull, efficient affair. Paley's watch would have been a large hunter or repeater enclosed within a silver or gold case. The back would hinge open to reveal a world in miniature of revolving gears in shining brass, tiny oscillating devices, and a great central spring, perhaps even (if a repeater) tiny bells to signal the quarter hours. The mechanics alone would qualify as a work of art as well as artifice. Paley's conclusion leaps from the page: the springs, gear wheels and all the other bits and pieces of the watch represent

11

the heart, the muscles, nerves, joints and all the thousands of tissues that constitute a living creature. If the watch and all its precisely interacting parts are made for a purpose, we are forced to conclude that we humans and all of living nature are also made to accord with a particular vision.

On the face of it, Paley's premise is incontrovertible. A watch could not have appeared out of thin air. It did not chip off something else, as the stone might have done. It did not grow out of the ground like the grass on the heathland. Any machine must be made; in the case of something so intricate as a watch, it requires a detailed plan and a craftsman of great skill. And, given that it has been made so carefully, even someone who had never seen a watch before would conclude that it must have some purpose, a function. Every machine of human invention, all its parts neatly functioning, precisely adapted to each other and to the whole, must exist for a purpose. (If you set out to make a machine that had no purpose, then *that* would then constitute its purpose.) From this it follows that the more we study natural organisms – the elegant 'machines' of nature – the more we will learn about and confirm the power, purpose and goodness of the Creator. Construed thus, science is an exercise in piety and the sciences dealing with life on earth provide an unending catalogue of arguments about the existence and benevolence of the Creator.

The concept of the body as a machine would have been as familiar to Paley's readers as the clockwork universe. A machine analogy for natural phenomena had long been a consistent element in Enlightenment philosophy. Leonardo da Vinci and the anatomists of the sixteenth century had seen the mechanical elements of a machine in the levers and pulleys of the skeleton and William Harvey, with the discovery of the circulation of the blood, made the body a dynamic mechanism. When Robert Boyle asked him what had inspired his discovery,

[Harvey] answer'd me, that when he took notice that the Valves in the Veins of so many sevral parts of the Body, were so plac'd that they gave free passage to the Blood towards the heart but oppos'd the passage of Veinal Blood to the Contrary: He was invited to imagine, that so Provident a Cause as nature had not so Plac'd so many valves without Design: and no Design seem'd more possible than that ... it should be Sent through the Arteries, and Return through the Veins[10].

Having given the outline of his case, on the very second page of his book Paley moved summarily to dispose of some easily anticipated criticisms.

Nor would it, I apprehend, weaken the conclusion, that we had never seen a watch made ... Neither, secondly, would it invalidate our conclusions, that the watch sometimes went wrong, or that it seldom went exactly right ... It is not necessary that a machine be made perfect, in order to shew with what design it was made ... Nor, thirdly, would it bring any uncertainty into the argument, if there were a few parts of the watch, concerning which we could not discover, or had not yet discovered, whether they conduced to the general effect.

These are the sorts of objections that any reader might raise, especially as, in 1802, the mysteries of reproduction were still just that, everyone fell ill at times (Paley was seriously ill as he wrote the book), scholars had long wondered about the uses of the apparently functionless appendix, and so on.

Paley was concerned with far deeper issues than this, however. He launched into a debate with all the leading 'atheistic' positions of the day. In Paley's typically eighteenth-century prose and sharp argumentation, the reader discovers a set of special issues and opponents. He fires off shots at a whole range of philosophers and philosophies, from Descartes to Locke, from Buffon to Erasmus Darwin. Their challenges to the foundations of Christian belief are both the immediate and eternal reason for his book. His special preoccupation is with what he considered the ultimate heresies – evolutionary theories. If Paley's arguments were framed in terms unfamiliar to modern readers it is because we have variously absorbed and discarded Locke on understanding and Cartesian/Epicurean atomism, and Buffon's and Erasmus Darwin's ideas about evolution were long ago rendered extinct by those of Charles Darwin.

William Paley devoted his life, and particularly his great skills as a thinker and writer, to God and will long be remembered as the man who set out nothing less than a proof of the 'existence and attributes of the Deity'. *Natural Theology* was Paley's last book, written in his old age. While even the most assertive or arrogant of intellectuals might have hesitated at attempting to produce a definitive proof of the existence of God – and Paley's book certainly shows itself as the work of someone who is very confident – in life he was modest and quiet, a somewhat shy, shambling figure, built short and square, and by 1800 suffering terrible pain from what was probably abdominal cancer. The only known portrait shows a man with a smile to light up even the most dreary northern winter day. His deep, quiet voice – always too soft for a properly dramatic manner at the pulpit – was the voice of calm and reason. As a student, he had never dreamt of adopting the citified modes of speech of Cambridge and London. His rough-hewn manner allowed him to speak directly to his flock and no one criticised or mocked his outmoded coat, his old-fashioned hat or his wrinkled stockings. At his prime, his sermons persuaded, cajoled and inspired rather than insisted or threatened. Those sermons, like his life, were full of wit and wisdom and coloured with just enough liberal thinking to make his superiors uneasy. As for inspiration, he never hesitated to build upon the works of others, indeed he mischievously advised young clerics: 'if your situation requires a sermon every Sunday, make one and steal five'. Towards the end of his life he walked with a broken, rolling, seaman's gait and would stop occasionally to recite aloud snatches of poetry or to sing.

He was born in 1743, the first child and only son of William and Elizabeth Clapham Paley. His father was headmaster of Giggleswick School and his mother a fine, intelligent woman, noted for her thrift. From them he inherited a flair for mathematics and a love of argument. Fifty-five years before Charles Darwin, he too entered Christ's College, Cambridge, where he graduated first of his class ('Senior Wrangler') in 1763 and then stayed on as Fellow of the college, teaching philosophy and the Greek Testament. From the very first a brilliant and much-loved, if unconventional, teacher, he was ordained in 1767. But the celibate life of the university don was not for him and in 1776 he married Jane Hewitt of Carlisle and became a clergyman and writer. His teaching at Cambridge had been so successful that his friends pushed him to publish his lectures, some of which had formed the basis of his first book, *The Principles of Moral and Political Philosophy* (1768).[11] This book alone would have earned him a place in history, enjoying some twenty editions in his own lifetime. *Morals* was followed by an influential study of St Paul and then, in 1794, Paley wrote a third book, arguably greater still. *A View of the Evidences of Christianity*[12] set out to show, conclusively and incontrovertibly, proofs of the historical truth of Jesus and the truth of his revelations. In this book, Paley put aside the exhortations of the pulpit in favour of the forensic techniques of the courtroom lawyer – a style that came very naturally to him. After graduating from Cambridge, he had for a while taught at a school in London.

15

There he spent his spare time equally at the theatre and the law courts. Both seem to have polished his rhetorical skills. He may even have been seriously tempted by the field of law and become a leading barrister at the Inns of Court; for his was a mind drawn naturally to logic, and to proofs and precision; instead he became a barrister for Christ.

In *Evidences*, he took the role of counsel for the prosecution, basing his case on evidence from four witnesses, the authors of the gospels. The lives, and not least the terrible deaths, of the Apostles and the other early martyred saints of the Church provided reason enough to believe, resoundingly answering the atheists' jibe: 'Either the Apostles could not write more intelligibly of the reputed Mysteries or they would not'.[13] Paley also insisted upon the authenticity of miracles as the vehicle for God's revelation of himself to man. They were an integral part of God's design and the essential mode by which God could communicate with the works of his Creation. 'Now in what way can revelation be made but by miracles? Consequently, in whatever degree it is probable or not very improbable that a revelation should be communicated to mankind at all, in the same degree it is probable or not very improbable that miracles should be wrought.'

Morals and *Evidences* brought Paley fame and a certain fortune and both became set books for examination of Cambridge students. (Remarkably enough, the last Cambridge students to be held responsible for the contents of *Evidences* sat warily down to their desks in 1920.)[14] But he never attained the bishopric that would have seemed the natural preferment for so revered a teacher and preacher. His highest appointment was Archdeacon of Carlisle. While his parishioners loved him, his peers may have found him just a shade too brilliant and too free with his 'almost too unlimited indulgence of wit and drollery' -- for example in that advice to young clerics over sermons.[15] In

Morals he had sided with Locke, whose work he had taught at Cambridge, over the right of people to revolt when their government failed in its responsibilities, a position he later abandoned in *Natural Theology*. And in one respect he was his own worst enemy – he refused to engage in 'rooting', his term for cosying up to people for influence. Nonetheless, the father of his great college friend John Law was Dr Edmund Law who, as Bishop of Carlisle, put a series of comfortable absentee livings his way, making sure that he had the security and contentment to write. The awkward man with his unfashionable accent and deep country manners was free to pour forth his brilliantly crafted texts, creating an intellectual achievement that has survived more than 200 years.

Evidences, with its insistence on the power of divine revelation, is obviously a mainstream Christian book, its goal to provide an independent line of support for the revelations that form the mainstay of Christian belief. It is often said of such books that their principal role is to comfort and confirm the believer rather than to persuade the atheist or sceptic, but in Paley's case this would be a cheap sneer. His technique was not to appeal to faith but to reason. All Paley's books are part of the maelstrom of ideas and movements that framed the Enlightenment and the Age of Reason. His 1802 masterpiece on God, through the strict logic of its author with all its strengths and flaws, visits the science of the age and the countervailing resistance of the natural world to simple arguments and neat solutions. Its full title is *Natural Theology: or Evidences of the Existence and Attributes of the Deity collected from the Appearances of Nature*. When, with the customary deference of the day, he dedicated his book to his bishop, Paley was at pains to point out that the work was intended to form a whole with his others, 'a system . . . the evidences of natural religion, the evidences of revealed religion, and an account of the duties that result from both'. In fact, in *Natural Theology* a different, more liberal, Paley emerges, carefully writing to persuade the deist or Christian alike. For an analytical reader like Darwin, the differences between *Evidences* and *Natural Theology* were striking.

The logical basis of Paley's argument would have been familiar to Darwin. The watch analogy was a syllogism, depending on the first two 'Rules of Reasoning' that Newton had laid down in his *Principia Mathematica* of 1687, the foundation stone of modern science: 'We are to admit of no more causes of natural things than such as are both true and sufficient to explain their appearances [and] . . . Like effects proceed from like causes.' Whether Paley can be thought truly to have abided by the first principle is debatable. What is true and sufficient is something to be determined, not taken for granted, in this debate. Newton's second rule gives greater support to the argument: 'Like effects [complexity] proceed from like causes [a maker].' However, the Scottish philosopher David Hume, among others, had already given the case against this kind of thinking:

> When we infer any particular cause from an effect, we must proportion the one to the other, and can never be allowed to ascribe to the cause any qualities, but what are exactly sufficient to cause the effect. And if we ascribe to it farther qualities, or affirm it capable of producing any other effect, we only indulge the licence of conjecture without reason or authority.[16]

Elsewhere he wrote: 'There can be no demonstrative arguments to prove that those instances of which we have no experience, resemble those, of which we have had experience.'[17] In other words, one cannot be confident in explaining what we do not know from what we do know, because we don't know what it is we don't know. Time and again we find the sciences, like other disciplines, exhibiting just this weakness, with false conclusions being drawn because of an incomplete vision of possible causes that in turn limits the imagination. It took Einstein, for example, to break the belief that light invariably travels in straight lines; he could conceive of something others could not (before it could be shown empirically).

The odd thing is that William Paley was not really a 'scientist' (a natural philosopher). He was not known as a naturalist, he did not collect insects or fossils as did so many of his colleagues, although he very much enjoyed angling. Although he had no training or experience in medicine, astronomy, chemistry or geology, the task he set himself was to turn the ploughshares of science into swords of religion. His dilemma, brilliantly resolved, was to find a way to use the contemporary fashion for rationality and science to make a case for God, when many scholars thought that philosophy and discovery were pointing in the opposite direction. He had not just to reconcile science and religion, but to use science to support, indeed to confirm, a belief in God; and not in some rearguard action, but a major offensive. For Paley, there was no luxury of time, however. Instead, there was a terrible urgency; he had to turn the scientists and philosophers against themselves before they could over-whelm his world. He had to affirm the existence of the Creator without getting caught up in contemporary arguments about biblical authority and the literal truth of every word of the book of Genesis. And he had to take on some of the greatest philosophers of the age.

Although they ended up on opposite sides of the issues of God, creation and life, Paley (in 1802) and Charles Darwin (starting around 1838) had to confront very similar problems. Both suffered the disadvantage of trying to make an incontro-vertible case without the kind of irrefutable empirical evidence

19

we usually describe as a 'smoking gun'. They had to convince
by argument because they could not 'prove', and therein lies a
restatement of Paley's dilemma: were his arguments founded
on scientific fact or pious belief? Were they the long-sought-after
proofs or only the familiar old assertions and appeals to faith?
Darwin, in turn, could describe natural selection but no one
had seen the origin of a new species actually happen. And for
both men, the growth of scientific explanations of material
phenomena conflicted directly with established beliefs and the
teaching of the Church. For Darwin, having at least started to
train for the Church, the burden of his discoveries was so great
that it made him a physical invalid. He knew the consequences
of his theory and the effect it would have on religion and thus
the very fabric of society. It would set people against each other;
it would set him against his own wife. If his theory proved
too revolutionary, it would be rejected out of hand. He would
become an outcast and all his efforts would be for nought. He
delayed publication for more than twenty years until he thought
the ground had been sufficiently prepared for his radical theory
of an evolutionary mechanism that would cut the intellectual
ground from under the feet of all the natural theologians.

Perhaps, then, there is a nice irony in the fact that when he
went up to Cambridge and reported to the porter's lodge just
inside that great gate, the young Charles Darwin was assigned
to the same rooms in Christ's College that Paley had lived in
seventy years before.

CHAPTER TWO

An Age of Science, An Age of Reason

'If we take in our hand any volume; of divinity or school metaphysics, for instance; let us ask, "Does it contain any abstract reasoning concerning quantity or number?" No. "Does it contain any experimental reasoning concerning matters of fact and existence?" No. Commit it then to the flames: for it can contain nothing but sophistry and illusion.'

David Hume, *An Enquiry Concerning Human Understanding*, 1748

'No man's knowledge can go beyond his experience.'

John Locke, *Essay Concerning Human Understanding*, 1690

Today we live – all too evidently sometimes – in an age of science. Science and its handmaiden, technology, shape every aspect of our lives. We might even envy people like Paley for having lived in much simpler times. But the turn of the nineteenth century was an immensely exciting time when both philosophy and science were stamping their mark on a broader cross-section of society than at any time since the Greeks. Already, the previous hundred years had been an age of discovery and experiment in everything from agriculture, blood transfusion and the discovery of oxygen, to inoculations against small pox, the first steam-powered carriages, and even calculating machines. People could now fly through the air in the Mongolfier brothers' hot-air balloons. Meanwhile, Britain's great mechanised mills (dark and Satanic) had begun to change the balance between countryside and town, agriculture and industry, self-sufficiency and reliance. In the process, both prosperity and poverty grew apace.

At its simplest, science (which in Paley's time was called natural philosophy) is an accumulation of wisdom and argument, facts and hypotheses, about what is. More fundamentally, science is about discovering causes: the why and how of the knowable world. Above all, science seeks explanations that can be expressed in terms of universal laws and therefore establishes a world of lawful, predictable behaviour. Sometimes we harbour the fallacy put about by scientists in the 1960s and 1970s that science (as expressed in today's extreme scientism) provides all the answers, and that it delivers certainty. Quite to the contrary, under science little stays the same. That is why it is so threatening to religious belief and socio-political authority. Science produces facts and laws but at its heart are questioning, testing and experiment, finding new explanations for old phenomena, finding new phenomena for old explanations, changing ideas and changing certainties. Religion, in contrast, is principally built upon certainties, authority and stability. 'A mighty fortress is our God' – a fortress against the surges of change that science and philosophy and, above all, independent thinking generate. Of course, 'religion', perhaps especially the Christian religion, is no monolith, any more than is 'science'. We use the words as shorthand for two kinds of intellectual and personal 'systems'. As a practice conducted by humans both may often fall short of the ideal and for the last 250 years they have been more opposed to each other than united.

In principle, science owes allegiance to no higher authority; as a wind of change, it bloweth where it listeth. Science is equally as dangerous for pointing out what is still unknown as it is for showing us new reliable facts. Science begets change and change always threatens the *status quo ante*, whether in rival fields within science or in religion. But orthodoxy, whether religious or political (or indeed scientific), depends upon commonly received opinions and often makes it heretical or treasonous to think otherwise. For all its innate conservatism, science always produces change. No scientist ever became famous for reporting that what we knew in 1870 or 1940 was best.

William Paley did not reveal what doubts he might have felt in the privacy of his study, but it seems unlikely that someone so well versed in science and so ready to do battle with the philosophical giants of his age could have failed to stare up at the stars in quiet moments with a niggling doubt about who else was out there. He would surely have pondered how to explain to his congregation that even something as reliable as the sun was not what it seemed. That the sun appears to orbit around the earth, disappearing each night and coming back up on the other side each morning, was one of the very first apparently reliable observations humans made about the universe we inhabit. It is far more 'obvious' than the notion that the earth is flat, for one can stand at the ocean-side and see that the horizon curves, and every sailor knows that when a ship appears from over the horizon, the tip of its mast shows before the hull. But nothing seemed more certain than the sun's movement and, unsurprisingly, the Bible is unequivocal about the fact that it 'goeth forth in his might' (Judges 5:30). That it was the sun moving, not the earth, was surely also explicit in the biblical story that, at Joshua's request, God made the sun stand still (Joshua 10:12–14). For Isaiah, God even made the sun move 'ten degrees backward' (II Kings 20:11).

Although some of his Greek contemporaries had doubts about the sun's movement, Aristotle – the great authority through the Middle Ages – had had none. He held two powerful theoretical positions about the geocentric cosmos: that the ideal shape was a sphere, and that the ideal motion was circular. From this he built up the view that the sun, moon and planets were each harnessed to a different revolving, perfect, crystal sphere, one inside the other, with the imperfect earth stationary at the centre. The ultimate expression of this system of spheres was in Ptolomey's *Almagest* or *The Great Syntaxis* (circa AD 160), on the strength of which Aristotelian cosmology reigned supreme for 1,500 years, until new astronomical calculations in the Renaissance, driven by the need for accurate, predictive star maps for navigation, began to force the creation of new explanatory models.

Nicolas Copernicus (1473–1543) in his *De Revolutionibus Orbium Coelestium* (published the year he died), forced the world to consider the heliocentric model in which not only does the earth revolve around the sun, but it also rotates on its own axis every 23 hours and 56 minutes. (Alternative models had proposed that, for example, the sun and planets stay still and the earth revolves, or that the sun and moon go round the earth and everything else goes around the sun.) Precise measurements made by Tycho Brahe (1546–1601) helped Johannes Kepler (1571–1630) make a new kind of astronomical sense. The movements of the planets – one of the great mysteries of the universe – could be boiled down to three very simple laws, all depending on the fact that their orbits were not circles but ellipses, all around the sun, except for the moon, which orbits the earth. The central consequence of Copernicus's revolution is only too obvious to us today. Not only had the earth been displaced from the centre of the universe, it had become merely a tiny speck of matter in the immensity of space, no more or less perfect than the rest.[18]

Copernicus died in 1543, leaving others to take up his work. Galileo Galilei was born in 1564 and acquired an immortal place in history for being forced by the Inquisition in 1633 to recant his belief in Copernicus's heliocentric universe. When he made his first telescope in 1609 and looked at the moon, he had started another revolution, discovering that it is not the ideal body that the ancient Greek philosophers had believed and every poet had romantically declaimed; instead, it was ugly and broken, pockmarked with craters and rifts. Perhaps even worse, he saw that the sun, which from time immemorial had been seen as a perfect sphere giving light and life to the earth, was also imperfect. In the Bible, it forms one of the great metaphors for the coming Messiah: 'The sun of righteousness shall rise with healing in his wings' (Malachi 4:2).[19] Galileo discovered that the sun is blemished, with dark spots that move, apparently randomly, about its face. (His opponents argued that since these spots could only be seen with the telescope they must have been artefacts of the lenses.) Shortly afterwards Galileo observed that Jupiter has its own planets and that Venus shows itself in phases, like our moon. He realised that, beyond the visible planets and fixed stars, there were millions of other stars, not visible to the naked eye, and who could guess what lay beyond those. In this new concept of the heavens, the earth and its inhabitants really were minor in significance. Perhaps there were even other worlds such as ours, with other sentient beings, and we were not alone in inhabiting the cosmos.

Although he was surely as complex a character as any other haunting the corridors of power and influence in seventeenth-century Italy, Galileo has become one of the more sympathetic characters in scientific history, an honest man cruelly oppressed by the enforcers of religious and intellectual orthodoxy.[20] Isaac Newton, on the other hand, presents much more of an enigma; a brilliant scientist whom we very much want to admire, but also a dark, brooding man, his personality seemingly pinched and bare. It is odd that such a profound intellect should have been so insecure about money and position, so temperamental

Galileo Galilei

and suspicious. While his ideas on motion have endured, his works on alchemy and vitalism have not. And he managed to gather about himself a host of eager rivals and competitors and suffered cruelly from the common problem that major discoveries seem to come in bursts, often being developed quite independently by more than one person at the same time. Many a schoolchild has damned Newton for inventing the accursed *calculus* (which he called 'fluxions') that he and Leibnitz invented independently. Newton was not helped by his great brooding sulks; he would start work on a subject, become dissatisfied because the answer didn't satisfy his own high standards, and then put it aside for a few years. Meanwhile others would be closing in on a solution.

England's greatest scientist, Newton was born in 1642, the year that Galileo died. He took over the science of Galileo's time and created a new intellectual sphere, a new world of mechanical laws with which we are still far more comfortable than we are with the modern world of quantum physics, where Einstein's relativity and Heisenberg's uncertainty reign. The ancients, particularly Aristotle, had thought that all matter was naturally at rest unless acted on by another force. This is again a weighty piece of common sense: when we see a stone on the ground, it does not move until we kick it. The question is, why does it then slow down and stop? Newton rewrote the science of motion and mechanics in a masterpiece of uncommon sense and mathematical precision.[21] He showed that all matter is in uniform motion (constant velocity, *including* a velocity of zero) unless acted on by an external force. Exactly opposite to Aristotle's view of motion, Newton showed that an object will remain still or continue to move at a constant speed in the same direction unless some external force changes things. A moving stone slows down because a force of friction has slowed it, not because it somehow wants naturally to come to rest. A thrown stone describes a parabola through the air, not because it naturally tends to progress in a perfect theoretical circle, but because a force – gravity – has diverted it from the direction in which we threw it. Single forces always act in straight lines, not circles. Any trajectory other than a straight line must be the result of multiple forces acting together.

One of Newton's most brilliant insights (with the assistance of Robert Hooke) was that the mechanism that keeps the planets in their elliptical orbits around the sun according to the rigid rules discovered by Kepler is the same as that which controls the fall of an apple from a tree, or shapes the trajectory of an arrow shot from a bow. From this he could predict that a projectile fired into the sky at a high enough velocity would continue indefinitely straight out into space. But one fired at some lower velocity would be attracted back to the earth by the opposing force of the earth's gravity, and if the two sets of forces *balanced*, then the projectile would settle into orbit around the earth – just as the earth and the other planets orbit the sun, and just as the moon and communication satellites orbit the earth.

Nothing would be the same after Newton's strict mathematics. Not even the simplest aspect of daily life on earth could be considered immune from the laws of science and the probing of scientists. Kepler's laws of planetary motion might be glossed over as remote and literally other-worldly but Newton's laws of motion touched every intimate detail of existence and were correspondingly subversive. Newton accelerated one of the great movements in science – which is to take the mystery out of everything. He also helped to explain some of the contemporary puzzles in the heliocentric model of the universe. If the earth is hurtling through space at many thousands of miles per second, why are we not all blasted off by the wind? If the earth revolves, and those in the northern hemisphere are standing up, why don't the upside-down Australians fall off? The answer is that gravity holds our atmosphere in place, so there is no cosmic gale, and it holds the Antipodeans in place too. For everyone, 'down' is towards the centre of the earth. Science had given nature a new uncommon sense. And, in addition to the technical importance of Newton's mathematics, the concept of a 'balance of forces' keeping the moon circling the earth and the earth in orbit around the sun, very quickly became a valuable metaphor for the description and explanation of a wide range of secular phenomena, including Malthus's ideas about population growth being held in check by negative factors and Darwin's ideas on evolution.

Newton's emphasis on matter and motion related centrally to the Epicurean school and their theories of the nature of matter itself. These ideas had been revised and extended in more modern times by the great French philosopher Descartes (René des Cartes, 1596–1650) whose physical theories Newton in turn largely supplanted. Beyond Galileo's collision with the Inquisition, if any one man could be said to have started the fields of science and religion on their course of conflict (or perhaps simply of divergence), it is Descartes. By sheer force of intellect and powerful original thought, he created a whole new approach to philosophy, brilliantly turning upside down the old, classical authorities to which the Church turned for support during the Middle Ages. Born in France and educated at the Jesuit college at La Fleche in Anjou, Descartes was Galileo's younger contemporary and a philosopher who wrote about everything from pure mathematics to human physiology, from the origins of the solar system to the fundamentals of human understanding. All his philosophy started with rejection of previous authority, none of which could be as reliable as one's own senses and intuition. Every schoolchild knows (or should know) his dictum: 'Cogito, ergo sum.' These three words, translated as 'I think, therefore I am,' represent his last resort after having rejected everything else in an attempt to find an incontrovertible reality – a truth – upon which to base a philosophical system.

The rigour of his methods was grounded first in mathematics: 'Those who are seeing the strict way of truth should not trouble themselves about any object concerning which they cannot have a certainty equal to arithmetical or geometrical demonstration.' Galileo, in one of his most famous passages, had put things even more eloquently: 'Philosophy is written in that great book which ever lies before our gaze – I mean the universe – but we cannot understand if we do not first learn the language and grasp the symbols in which it is written. The book is written in the mathematical language ... without the help of which it is impossible to conceive a single word of it, and without which one wanders in vain through a dark labyrinth.'[22]

Before Darwin

As a young man Descartes wandered the capitals of Europe before settling in Holland in 1628. From the beginning he thought intensely about epistemology: the question of how we know, and especially how we can find ultimate, objective ways of knowing what is right and true. Again this turns on his basic premise, 'Cogito, ergo sum.' In his *Meditations* he turned this into a long argument for why God must exist, why God is perfect, and why God has made man in his own image. In considering the workings of the human body, he drew a firm line between animals and ourselves. Humans alone have a dual nature – a material body and an immaterial soul – and that distinguishes us from the rest of creation. This was a distinction that carried far into the nineteenth century even as people began to discover the workings of the nerve impulse and the brain and as they delved into the nature of consciousness; until they began to find that the old line between animals and humans was as blurred in this regard as in every other. Descartes' physics of the universe was based on the idea that the planets were suspended in a total void and that their motions described a series of vortices. Like Newton's mathematics of forces acting in straight lines, which replaced them as a description of the cosmos, vortices had a strong metaphorical as well as actual ring to them. But Descartes did not believe that bodies could influence each other except when in contact. By dismissing 'action at a distance', and therefore phenomena such as gravity, while having moved ideas forward mightily, he failed to create a truly modern physics.

In his cosmology, Descartes began with an Epicurean physics, seeing the world arising out of atoms in motion. Democritos (*c.* 460–371 BC), in perhaps one of the most prescient pieces of pure intellect, had taught that: 'Nothing exists except atoms and empty space; everything else is opinions.' In this atomic theory, all the various kinds of matter differ only in the size, shape and motion of their infinitesimally small atomic constituents. In Descartes' atomic-deist theories, creation was originally a series of events during which order condensed out of this random atomic behaviour. All matter – whether rocks, trees or monkeys – is merely the combinations of these atoms churning through space, driven by chance. God then had been relegated to the maker of the atoms and the formulator of the broad rules of their motion. By Paley's time, despite the failure of the theory of vortices, Descartes was popular with deist scientists trying to find new truths about the cosmos in the spirit of the Enlightenment and Age of Reason, but was dismissed by traditional theologians as one of the 'ancient sceptics who have nothing to set against a designing Deity, but the obscure omnipotency of chance, and the experimental combinations of a chaos of restless atoms'.

In parallel with such philosophical approaches to knowledge itself and new theories about the very state of matter, a fresh style of experimental science flowered in the modern intellectual environment. Deep thought and practical experimentation fed off each other; as one scholar probed into *how* we know, another tinkered with new devices for observation and discovery. One man in particular helped launch this empirical renaissance. Francis Bacon (1561–1626), in his *Advancement of Learning* of 1605 and *Novum Organum* (1620), laid out a template for science to proceed by the accumulation of facts and by the framing of rational, testable hypotheses. This empirical approach was based on the revolutionary notion that truths about the material world should be discovered rationally through experiment, observation and analysis rather than derived from a set of classical philosophical abstractions or presented as a matter of divine revelation. In *Novum Organum* he wrote: 'There are and can be only two ways of searching into and discovering truth. The one flies from the senses and

31

Robert Boyle

particulars to the most general axiom ... The other derives axioms from the senses and particulars, rising by a gradual and unbroken ascent, so that it arrives at the most general axioms last of all. This is the true way, but as yet untried.'

Perhaps nothing better exemplifies the new spirit of empiricism that flourished in the second half of the seventeenth century than the experiments of Robert Boyle and his colleague Robert Hooke, two of the most brilliant natural philosophers of their age, who worked together as equals but in origins and personal style were as different as they could be. This was the first generation of natural philosophers who could be considered 'scientists' as we understand the word. The Honourable Robert Boyle was the wealthy son of the even more wealthy 1st Earl of Cork; Hooke came from a family of more modest means: his father was a parson who died young. Boyle, educated at Eton, did not attend university. From an early age he had been an avid reader and after schooling at Eton was tutored privately, first in England and then, from the age of fourteen, in Geneva. Back in England at eighteen, he took up chemistry and then settled in Oxford where he built a laboratory and hired the young Hooke to assist him. In pictures painted in his middle age, he looks a magnificent rich dandy, tall, haughty and remote, but in reality he was a frail man, often ill, with a stammer and a mild, kind, generous and refined intellect, to whom many potential honours, including a peerage, were offered. After leaving Oxford for London, in part to take a greater role in the Royal Society, his intellectual interests ranged well beyond the laboratory to philosophy and particularly to the promotion of religion. If anyone of his age had the right to be called a true philosopher of nature, it was Boyle. He never married but lived most of his adult life with his sister Lady Ranelagh. When she died in 1691, he died just a week later.

Robert Hooke, born in 1627, was eight years younger than Boyle. He was born on the Isle of Wight, where his father was vicar of Freshwater. He soon showed an advanced ability in drawing and everything mechanical, but had to make his own way in the world at thirteen, following the death of his father. A small inheritance allowed him to become an apprentice to the painter Sir Peter Lely in London but Hooke soon decided that he had enough skill in that direction without the drudgery of apprenticeship. He entered Westminster School, where he demonstrated an amazing ability to master languages, learnt 'the six books of Euclid in one week, mastered the organ in twenty lessons, and invented thirty ways of flying'.[23] He became an undergraduate at Christ Church College, Oxford, in 1653 where, essentially penniless, he was forced to earn his way as a servant to another student. Nevertheless, he soon made his

Robert Hooke

abilities known to all the scientific luminaries of the age including Christopher Wren and Robert Boyle.

Boyle and Hooke made a superb team, with complementary skills and an equal commitment to the new-fangled Baconian idea that the truth could be found through direct observation and experiment.[24] During the twelve years they worked together at Oxford between 1656 and 1668, Boyle and Hooke's ideas led in every direction. They became particularly famous for investigating the properties of air, which in classical and medieval times was one of the four 'elements' (air, earth, fire and water). Using their own version of the air pump that had been invented by Otto von Guernicke, they measured the elasticity of air and found the mathematically precise, inverse relationship between the volume and pressure of a body of air. This is one of the first physical laws to be enunciated and is still known today as Boyle's law. Boyle's air pump, built and operated by Hooke, was also intended to show that Aristotle was wrong when he taught that a vacuum was impossible in nature. Attached to their pump (which often broke down) was a glass chamber inside which they could create both high pressure and a (partial) vacuum. They proved that when the air was removed from the chamber, sound could not be transmitted, although light could. A whole mini-revolution was seeded by what might seem to us a very simple piece of apparatus when they also used it to conduct some elementary experiments on the effect of the air on living organisms, putting a bird or mouse into the chamber and evacuating the air. As it was withdrawn, the animal became listless; if the air was restored, it revived. If enough air was withdrawn, the creature died: all commonplace stuff to us, but revolutionary then. They had discovered that there must be some 'vital essence' in the air that makes life possible. This, and the eventual discovery of oxygen by Lavoisier, Joseph Priestley and others, launched an investigation into the material (physiological) basis of life itself, a subject with enormous metaphysical implications given that it had always been thought that life was something breathed into creatures by God, not just another property of matter in motion.

Linguist, microscopist, artist, mathematician, mechanical experimenter and inventor, palaeontologist, surveyor – Hooke's accomplishments were long overshadowed by the fame of the two other geniuses (Boyle and Christopher Wren) with whom he worked. With Boyle he was the master experimenter and inventor – for example of the universal joint, essential to so many modern machines.[25] With Wren he was the great engineer. When Wren was made architect for the rebuilding of London after the Great Fire of 1666, Hooke was the chief surveyor; and as Hooke the engineer he gave Wren the parabolic formula for the great dome of St Paul's Cathedral.[26]

For a period after Boyle left Oxford, he continued to employ Hooke, who became the 'curator of experiments' for the Royal Society and therefore found himself (or made himself) for the rest of his life at the centre of every scientific discovery of the age. Sadly, few people seem really to have liked Hooke, whose

35

childhood kyphosis steadily worsened so that as an adult he became a twisted hunchback. Something of a miser and a misanthrope, and never one to avoid a fight or to allow someone else to take credit for his own discoveries, he became extremely litigious. He contested bitterly with Christian Huygens over the invention of the spring-regulated mechanism that made a pocket watch (and thus Paley's famous metaphor) possible, and he argued bitterly with Newton over optics and cosmology. When Newton took over from Hooke as President of the Royal Society, Hooke's portrait mysteriously disappeared from the society's rooms.

The combination of Newtonian mechanics, Baconian methods, and the new experimental empiricism that propelled eighteenth-century science was also reflected in the Industrial Revolution and is exemplified in the extraordinary career of William Paley's contemporary, Erasmus Darwin (1731–1802). This Darwin, grandfather of Charles, was a successful doctor in the city of Lichfield and at the same time a member of the famous Lunar Society of the industrial Midlands. Darwin, Watt, Boulton, Wedgwood – all the great names of British inventiveness – met once a month and created a new kind of intellectual centre outside the universities.[27]

Erasmus Darwin was not just a brilliant doctor, a trencherman (he weighed some eighteen stone) and sensualist (fathering fourteen children, at least two out of wedlock), he was also a remarkable inventor. Designs for steering mechanisms and sprung wheels for carriages, a steam-driven carriage, improvements to steam engines, a horizontal windmill, the canal lift, hydrogen balloons to carry mail, a clockwork-driven artificial bird, a copying machine, a turbine engine, a multi-mirrored telescope, a water closet, devices for improving gardening, new kinds of spinning machines – all flowed from his pen. Further-

Erasmus Darwin

more, he was also a poet and radical philosopher who used his poetry, tending to the epic in style and volume, as the vehicle for his most dangerous ideas.

One of the themes of Erasmus Darwin's immensely popular works concerned what we now call evolution, a subject more commonly associated with his grandson. Like so many of his century, Erasmus Darwin was fascinated by fossils and the extraordinary record they presented of life and death over the ages. Darwin saw that they were evidence of the life and death of legions of organisms never seen alive by man: extinct forms about which the Bible is totally silent. Fossils were evidence of a whole ancient world waiting for scientific explanation. Without knowing how much time might have been involved, and ignoring the biblical narrative of creation, Erasmus Darwin proclaimed a world of gradual change over the aeons; change from simple creatures to more complex; life arising out of chemistry, driven by the forces of the environment:

> Earths from each sun with quick explosion burst,
> And second planets issued from the first.
> Then, whilst the sea at their coeval birth,
> Surge over surge, involv'd the shoreless earth, Nurs'd by warm
>   sun-beams in primeval caves,
> Organic life began beneath the waves . . .
> Hence without parent by spontaneous birth
> Rise the first specks of animated earth.[28]

'Without parent'! No amount of argument could make that idea compatible with 'God the father' and the most honoured of all words in the Old Testament, the first verses in Genesis, which state that God created the world exactly as we know it, in six days. Erasmus Darwin had issued a challenge in the style of a dictum of Descartes, who had once said: 'The nature of physical things is much more easily conceived when they are beheld coming gradually into existence, than when they are only considered as produced at once in a finished and perfect state.' Other scholars in France and England shared this vision of a changing world, but 'gradual' and 'chemistry' were not in the Church lexicon. Paley read Erasmus Darwin, recoiled, and reached for his pen.

Even more threatening to Paley's world view was the quickly growing sciences of the earth, brilliantly synthesised by James Hutton, a Scottish doctor, farmer, philosopher and geologist who, in 1795, published a two-volume *Theory of the Earth*.[29] If any single book captured the challenges posed by the new science, it was this. Genesis says that the physical world was created in three days and populated by animals and plants by the sixth. Learned clerics had even devised elaborate schemes to decode the histories recorded in Genesis to arrive at a date for this great event – 4004 BC. But dozens of equally learned men who had been investigating the nature of the earth itself had produced a different kind of authority in new empirical data as well as theory. Hutton distilled the results of a hundred and fifty years' enquiry into the structure of the earth and the processes that shaped it, and dared to suggest a totally different conclusion: that the world was unknowably old.

In fact, the possibility of an ancient earth had been proposed and dismissed many times before Hutton, even by Aristotle. A preoccupation of many eighteenth-century writers about the earth had been to counter theories, like Aristotle's, of an eternal earth having neither a beginning nor an end. The authority of Genesis must be greater, they insisted: the world must have had a Beginning and was proceeding to a definite End. But Hutton supported his new ideas both with solid empirical evidence and an underlying theory based on a Newtonian balance of forces. He saw a pattern in the history of the rocks: gradually worn down by erosion, washed into the seas, accumulating as sediments, raised up as new dry land, only to be eroded again. Not the linear narrative of Creation to Final Conflagration that the Bible foretells, but something cyclic, balanced, timeless, unending. Hutton also openly espoused Erasmus Darwin's ideas about organic change; they helped explain the successions of life that had inhabited his recycling globe. And the threat that Hutton's geological science posed was the greater because, where Erasmus Darwin had a wild-eyed hypothesis, he had a cold, sober theory.

The strength of Hutton's case made science and a literal interpretation of biblical creation virtually irreconcilable. Now too many of the central, commonsensical dogmas of religion had been replaced by theories that were not just difficult to understand and to prove, but also challenged the central core of established belief. Little wonder, then, that someone was needed to respond to all these challenges from the side of organised religion.

Writing in 1802, however, Paley had greater challenges to face even than these. While natural philosophy was concerned with the definition, description and material causes of natural phenomena, Paley also had to engage with moral philosophy, which is concerned with values, meaning and purposes, and metaphysics, which probes the ultimate nature of what is, and how we know. Science is all about causes: what causes the apple to fall from the tree? What caused the apple to become separated from the tree so that it could fall? What caused the apple to ripen? What caused the apple tree to flower and the bee to pollinate it? What caused the tree? But in a religious context, cause and its metaphysical counterpart 'purpose', look very different. For a theologian in Paley's time, as today, God was always referred to as the First Cause. Then there are Second Causes, which are due to the inherent nature of matter and material systems and the operation of natural laws. Newton's laws of motion and gravitation, Kepler's laws of celestial mechanics, the laws of thermodynamics, gravity, the laws of chemistry (for example, the valency of atoms), and (in our time) the coded and coding sequences of amino acids in the DNA molecule, are all formal expressions of Second Causes. In turn they depend on further nested sets of causes in nuclear, atomic and quantum mechanics, and so on. In a sense, all of science – which is a system of investigation based wholly in material properties and processes – is the discovery of Second Causes.

If these Second Causes shape and drive the daily economy of the earth and its cycles of life and death, we are presented with a dichotomy. Now there are two rival views of God: one (more or less the Christian God) is a creator who is also the endless, continuous, loving God who has counted every hair on our heads and sees the fall of every sparrow. This God not only created but continually *directs* his Second Causes. The daily business of the world matters to this God, and particularly to his son Jesus. The alternative is a more distant (deist's) God who created the world and then set it to run like a giant cosmological train set or clock. All its processes and phenomena – orderly and predictable, contingent and occasionally random, from the movements of the heavens to the physiological bases of diabetes, and by which the cosmos has changed over millions of years – all these are the results of Second Causes, flowing inevitably from the nature of matter itself. In this case, once God had created the conditions for Second Causes and the rules of their operation, he made it unnecessary to have a direct hand in every single act of man and nature.

The dilemma created by the new scientific philosophies was therefore the potential relegation of God from all-powerful to first power only, and the acknowledgement that other scientific (Second) causes drove the world day by day, year by year. The consequent and even greater dilemma was that, once one admitted Second Causes, it was only a simple extrapolation to *all* the processes of life being definable in terms of such causes. In the process, the need for a First Cause would simply fade away. There would be no room – no need – for God at all. All causation might ultimately be, as Erasmus Darwin put it, 'without parent', nothing more than the result of chance collisions of atoms in empty space, as proposed first by Democritos and the Epicureans. Indeed, Descartes had even suggested that, if one could know the nature and precise motion of all the atoms in the universe at a single moment, one could predict their future arrangements; in other words, one could predict the future (Although if the future were simply the inevitable extrapolation of the present movements of atoms, it would also mean there was no such thing as free will).

41

Beyond First and Second Causes, there is the Final Cause –
the purpose that God purportedly had in having created the
world, and the end goal of all its daily operations, summed up
over the millennia. Traditional Christians, with their emphasis
on the Trinity, on Revelation, and on the promise of Redemption, naturally believe in the concept of Final Cause – purpose.
Putting it simply, God has in mind a purpose for each of us
and for the whole world he created. That is why, even though
Second Causes may be operating, he still steers the ship. This
is not a God who has set the world going like some autonomous
machine; the Christian's God is one who will eventually,
through his Son, redeem all our sins. First Cause, Second Cause,
Final Cause – all (relatively) easy to believe in, difficult to live
up to, and hard to prove.

If there is no God, then there is no purpose. And the reverse
might be true: if there is no purpose, there is no God. In Paley's
time, the 'death of God' was a very distant, if still fearful,
prospect. God's role as First Cause and Final Cause was, for
the moment, reasonably secure, even among the most radical
of philosophers such as Descartes. As Robert Boyle wrote in an
early classic essay:

> Epicurus, and most of his Followers . . . Banish the Consideration
> of the Ends of Things; because the world being, according to
> them, made by Chance, no Ends of Things can be suppos'd to
> have been intended. And, on the contrary, <u>Monsieur Des Cartes</u>,
> and most of his followers, supposed the Ends of God in Things
> Corporeal to be so Sublime, that 'twere Presumption in Man to
> think his Reason can extend to Discover them.[30]

But there is a telling passage in the work just quoted. It reveals
what a scientist sees as an unacceptable inconsistency but a
theologian sees as evidence of God's omnipotence. Boyle had

to allow: 'Nor is this Doctrine [of Final Cause] inconsistent with the belief of any True Miracle; for it supposes the ordinary and settled course of nature to be maintained [while] the most Free and powerful Author of Nature is able, whenever he thinks fit, to Suspend, Alter, or Contradict those Laws of Motion, which He alone at first establish'd and which need his perpetual Concourse to be upheld.' This seems to be playing both ends against the middle. If God can do anything he likes (as in causing the sun to move backwards), why has he bothered to set nature going as a system according to the strict laws he gave it? In science, the exception never proves the rule.

Final Cause is rather difficult for a scientist to come to grips with. It is rather like the soul, belonging outside the material and the natural. Science is particularly the opposite of everything supernatural; its very nature is to challenge and explain away anything that smacks of a paranormal world. At its harshest, science would deny that such realms exist; at its most charitable it would say that such things must be studied by some other logic. In any case, whatever a reader in 1800 thought of 'purpose' in nature and in philosophy, the cosmologists had long since reset the terms of intellectual engagement: the universe now had to be seen as moving according to precise laws; Newton had reduced the laws of cosmology so brilliantly expounded by Copernicus, Kepler and Galileo to the laws of everyday existence. The moon is held in orbit around the earth, and the earth (and moon) around the sun, by the same force that causes the apple to drop four feet from the tree. In the process, natural philosophy had produced some conflicting and dangerous new truths and consequences. On the one hand, the cosmos might have been created by God in an instant of time, remaining fixed and unchanging until he sees the need to destroy it and to redeem his people. On the other hand, the world might be imagined as coalescing out of a whirling mass of fiery atoms,

43

now cooling. The latter world was one of continuous change rather than the traditional series of fixed points along a pre-ordained path: Beginning, Fall, Deluge, Final Conflagration. In such a world of material causes, people had – if they dared look – a new view of free will, predestination and, perhaps more subversive than anything else, change. No wonder most societies had previously kept literacy and learning for a very few, with only the controlling seers being allowed to contemplate the eternal mysteries. For ordinary mortals (to paraphrase), a little knowledge would be an extremely dangerous thing. Once Pandora's box was opened and a new, lesser, role ascribed to God, who could predict where matters would end?

CHAPTER THREE

Problems at Home

'There are more things in heaven and earth . . . than are dreamt of in your philosophy.'

Shakespeare, *Hamlet*, I, 5

If the Church found itself besieged by the discoveries of science, it found little support from metaphysics either, starting with Descartes, whose philosophy demanded new rigour and personal judgement in the search for proofs of what we know. In such a philosophy, mysteries like faith and revelation are unreliable guides to the truth. A very similar line was argued by John Locke (1632–1704) who, in his *Essay on Human Understanding*, also explored the very nature of knowledge. To what extent does reality exist outside of our perception of it? Knowledge involves the relations of ideas, but ideas do not exist outside of the mind and experience. Does the mind therefore contribute to the reality of things, or in fact remove them further from reality? In fact, Locke becomes rather confusing on the subject of matter, or substance, which in philosophical terms was recognised through its three essential properties of solidity, extension (property of occupying space) and action (motion). Bishop Berkeley (1685–1753), on the other hand, in his 'immaterial hypothesis', was quite emphatic on this subject: 'No matter exists except in our perception.' Other than our own persons, 'all other things are not so much existences as manners of the existence of persons'. The great Dr Johnson (1709–1784),

John Locke

predictably enough, thought that the idea of the non-existence of matter was nonsense. 'I refute it thus,' he exclaimed, kicking at a stone (and giving it a distinctly Newtonian acceleration).

When John Locke taught that 'Reason must be our last judge and guide in everything,' he (along with continental philosophers such as Spinoza, Leibnitz and Gassendi) put his seal on the Age of Reason, and at the same time laid down a challenge. His philosophical system has no room for blind faith. 'Faith is nothing but a firm assent of the mind: which, if it be regulated, as is our duty, cannot be afforded to anything but upon good reason ... He that believes without having any reason for believing, may be in love with his own fancies; but neither seeks truth as he ought, nor pays the obedience due to his maker.' In that case, a bishop's (or a pope's) say-so was definitely not a sufficient reason for believing anything. And nowhere was any philosophy more threatening than when it turned a hard-edged logic and the disciplined, unforgiving eye of reason onto the Bible itself, and particularly onto the miracles

– those supernatural episodes in the life of Christ and the saints that form the very basis of Christian revelation. As Paley insisted in *Evidences*, for the Church the miracles are God's way of vouchsafing to his people the authority of his purposes for them. Otherwise, God is materially unknowable; in all normal respects the Holiest of the Holy Ones is hidden from the people. The 'breath' and 'hand' of God are figurative, not literal. The special exceptions take the form of paranormal phenomena – a burning bush, or the sun standing still or moving backwards, for example. Later, he sent his Son to save the world, and the miracles of the New Testament (starting with the immaculate conception – 'for behold a Virgin shall conceive and bear an son' – and proceeding through the raising of Lazarus from the dead, the feeding of the five thousand, and ending with the Resurrection itself) are his way of demonstrating the Saviour's bona fides. But throughout the eighteenth century, philosophers had secretly or openly questioned the reality of miracles, requiring that the miracles of Christ either be explained in material, scientifically understandable, terms or rejected (the raising of Lazarus, for example, looks exactly like an example of cardio-pulmonary resuscitation).

One of the great joys of the Renaissance, Reformation and Enlightenment was the freedom philosophers, scholars and ordinary people acquired to think for themselves. Hence the search for proofs of the unprovable, for the qualities of the ineffable, and for facts about the unknowable. Today we take this freedom for granted, and equally cheerfully slide over some deep philosophical difficulties. Not so for the Scottish philosopher David Hume, whose own epistemology has been no less influential than that of Descartes and Locke in shaping Western thought. Much of Hume's philosophical writings are necessarily deep and abstract, but others are set in more familiar terms and deal with readily appreciated (if still dangerous) questions.

Hume was born in Edinburgh in 1711 and had early trained for the law – a subject that appealed to him about as much as medicine at Edinburgh later attracted Charles Darwin. His passion was for a wide learning based in philosophy, starting with the classics of Cicero and Virgil. He had been a delicate young man, in large part because of his intense studying and, evidently, hypochondria. To help recover from a depression or mental breakdown, he travelled to France, living as frugally as possible, and eventually spent two years at La Fleche in Anjou where he had access to the library of the Jesuit college. It was here that he wrote his monumental work, *A Treatise of Human Nature* (published between 1739 and 1740). At the college, the same institution where Descartes had studied, he heard a Jesuit teacher explaining a recent 'miracle' and immediately wrote out a rebuttal of the whole concept. For Hume, the miracles reported in the New Testament were 'a contest of opposites . . . that is to say, a question whether it be more impossible that the miracle be true, or the testimony real'. For its day, that was almost as heretical as the present-day argument that some, if not all, of the miracles are mere fictions, constructed as a mythology around which to unite the fledgling first-century Christian Church.

Hume aimed to elevate moral philosophy to a science following the examples of Bacon and Newton in natural philosophy. A sceptic, he lost his faith very early, perhaps when a student at Edinburgh. The *Treatise of Human Nature* was the foundation of his later fame, even though the book earned little for him at first; in fact it was a commercial failure and a bitter disappointment, Hume wryly complaining that it 'fell still-borne from the press'. Needing to earn a better living than his pen afforded, he spent a curious period first as a tutor to the insane son of the Marquis of Annandale and then as secretary and

David Hume

judge-advocate to General St Clair on an expedition against the French at Port Lorient, Guernsey, at the end of the War of the Austrian Succession. While now relatively hale and hearty, the life of a man of action scarcely suited him, but attempts to procure a more suitable position, a chair at Glasgow then one at Edinburgh, failed.

Following the death of his mother he returned to the family home in Scotland and began a period of great productivity, during which, in 1751, he wrote two works that had special bearing on an evolving, perhaps revolutionised, view of religion. In his *Natural History of Religion* (1757) and *Dialogues Concerning Natural Religion* (only published after his death in 1779)[31] he used all his powers of reason and argument to test the case for traditional revealed religion as opposed to the deist

49

position, that everything that was important about religion could be (must be) derived by reason alone. In the *Dialogues* Hume carefully covered his tracks by laying out his arguments in the form of a conversation among three different philosophers. This was the familiar philosopher's device that Galileo used in his *Dialogue concerning the two major world systems* (1632) – without fooling the Inquisition. Both Galileo and Hume were following the example of Cicero's *De Natura Deorum*, or *On the Nature of the Gods* (77 BC), which is set as a debate among Epicurean, Stoic and Academic philosophers.

Hume's *Dialogues* did its damage not so much in the conclusions it reached as in daring to ask the awkward questions: What do we know that is true and independently verifiable about God, as opposed to what we are told? How do we know? Hume even asked the evolutionary question: Is it not more logical to assume that complex living creatures had their origins in simpler ones than via some miraculous creation by an infinitely powerful, but nonetheless unknowable, designing intelligence? Hume challenged everyone who thought they could find a rational basis for understanding God. He expressed the problem very simply: If we had never thought about there being a God in the first place, would objective, rational investigation and argument necessarily uncover his existence and define his nature? If we are already sure that God exists, it is not difficult to find seemingly rational arguments to support that notion and even to conclude that God must have certain specifiable 'properties', but if we were to start from scratch – if we were immigrants from some distant pagan shore or outer space – would objective study of the natural world and deep philosophical enquiry produce ineluctable proofs of God? Would there be miracles and signs, for instance, that could not be explained away rationally? Was Voltaire right when he said that if God did not exist, it would be necessary to invent him?

And we always have to worry about the final ace that Hume has up his sleeve: logically, he says, anything that can be imagined as existing can also be imagined as not existing. For every piece of evidence we can find for the Creator, we have to allow the existence of equally powerful evidence against. Hence Paley's dilemma.

As for religion itself, it would be wrong to think of it as having been a passive spectator at these feasts of the intellect. Indeed, the Church and the churches became their own best and worst friends. Ever since Martin Luther in 1517 nailed his Ninety-five Theses to the church door in Wittenberg and unleashed a flood of independent thinking about forms of worship and modes of belief, it had become impossible for the Church to speak with one voice and proclaim one doctrine. Instead, many voices, doctrines and practices competed for people's attention. A single original discipline had been opposed by a structureless freedom reaching to the heart of belief. Each group was defined on nuances of doctrine and separate routes to personal salvation, defended in the name of reason.

One such argument was between the theists and deists over the critical matter of revelation. For the Christian Church, revelation meant the events, especially the miracles, by which God had communicated with his chosen people, and it especially meant God's self-revelation in the form of his son Jesus, sent for the redemption of our souls. Deists, however, insisted that revelation was just the public-relations machinery of a controlling priesthood. God was enough to stand on his own, with the vast panoply of nature itself forming the only necessary evidence of his Being. In the Age of Reason, therefore, rational study of the world alone could reveal the Unworldly One. As Thomas Paine (of American Revolution fame) was to write in his deist manifesto:

When the divine gift of reason begins to expand itself in the mind and calls man to reflection, he then reads and contemplates God and his works, not in the books pretending to revelation ... The little and paltry, often obscene, tales of the bible sink into wretchedness when put in comparison with this mighty work. The deist needs none of those tricks and shows called miracles to confirm his father, for what can be a greater miracle than creation itself, and his own existence.[32]

Trinitarians opposed Unitarians, Arians and dozens of other sects over the divinity of Christ and the identity of the Holy Ghost and 'Christ with God'. Was Christ the same as God? (Sabellians); was God different from the Holy Ghost and again from Christ, who was his son? (Trinitarians); or was Christ (as the Socinians argued) merely another in the line of prophets? Then the Sub- and Infralapsarians opposed the Supralapsarians on the question of whether the Fall of Man was intended by God or only permitted after he saw man's wickedness – an argument that parallels the dispute among the Armenians, Calvinists and others over the issue of the predestination of individual salvation. Many English sects dissented from the Thirty-nine Articles that defined the core precepts of the Church of England. Among them were the Occasional Conformists and Non-Jurant Schismatics, who otherwise remained true to the doctrines of the Church but rejected part or all of its discipline. And there were those who dissented against practices of worship and inclusion, the latter including the Baptists, Anabaptists and Paedobaptists, with their differing views on the issue of baptism and consent.

After 1662, non-conformists of every stripe in England were persecuted with a new and ruthless zeal, but they always bounced back. Many schismatic sects, such as the Plymouth Brethren, became even more rigid in matters of piety than the

John Toland

Church of England or Catholics from whom they had split. Others were quite liberal in the interpretations of the Bible, particularly the Mosaic account of creation, thus allowing their followers to reconcile the new discoveries of science with their beliefs. As long as anyone insisted that Genesis remained the one unimpeachable source, however, the obvious result was confrontation and discord.

In 1696, all Europe was scandalised by the radical scepticism of John Toland, whose book *Christianity not Mysterious*[33] was ordered to be burned by the public hangman in Ireland. Toland was born in Ireland and brought up as a Catholic, then became a Protestant and a free thinking rebel, eking out a living as a writer of highly polemical tracts and books and dodging from country to country just ahead of a host of would-be persecutors. (Among other accomplishments, he invented the term 'pantheism'.) His writings exemplify what happened when free thinking and (even more dangerous) outspoken populists started to apply the pure reason of thinkers such as John Locke to a close study of the Bible. Toland thundered: 'Whatever is contrary to Reason can be no Miracle, for it has been sufficiently prov'd already, that Contradiction is only another word for Impossible or Nothing.' Toland dared to write what many felt, that it was absurd that the wine at the communion service should be thought literally to be transubstantiated into the blood of Christ. It was absurd that the disciples could have seen Christ walk on water. If you can believe in miracles, Toland argued, what is to prevent you from believing any nonsensical fiction? A church that depended on subjecting its adherents to the discipline of believing in miracles, and held its members in awe of the unknowable, was not worth belonging to. The precepts of the Church had to be understandable in material terms and expressed in plain words.

A hundred years later, the legacy of this free thinking made for a particularly dangerous time for the established Church of England. The Church was part and parcel, warp and weft, of the oligarchy; any threat to it threatened the very fabric of society. We must also remember that in 1802, Britain was at war with France. The threats from across the English Channel were not just the liberal intellectual challenges of the free thinking French Enlightenment, from Descartes and Buffon to Rousseau and Condorcet, but also the political challenges of the French Revolution, the material horrors of the Terror, and now the wars being waged by Napoleon. Riot and revolution, free thinking and self-improvement, tyranny, war and savagery were everywhere. One would readily be forgiven for wondering whether all this modernity was a good thing.

Paley therefore did not set out to write his proof of the existence and attributes of God in a world of certainty. There were enemies from without to be countered: materialist and rationalist enemies of the ineffable, scientists and philosophers from Britain and the Continent. And there were enemies from within: religion was beset by complex philosophical debates that threatened the whole basis of belief. Throughout it all, God's purpose was becoming harder to read, certainly more difficult to proclaim. At the beginning of the Age of Enlightenment the problem had been to find a secure place for science in a religious world; by the end, the problem was exactly the opposite: if the world operates through Second Causes, where was the role of God? One solution was to insist on the literal truth of the biblical story of creation: but that necessarily represented a denial of the discoveries of science about the age of the earth (and universe) and the role of change.

Two issues, above all others, motivated William Paley: the biting scepticism of the philosophers John Locke and David Hume, and the nagging threat of a theory of matter consisting of space and atoms in random motion. By 1800 such theories had long since spawned versions of the ultimate atheism: evolution. Scepticism could be countered with logical argument, but a rival explanatory theory – especially a godless theory like atomism – was an even greater threat. We can measure the challenge that a self-ordering world, operating on independent laws and motions – and, above all, on chance – posed to received religion by the bitter rhetoric of the defenders of the orthodox. We can gauge how long-standing this threat had been – since Descartes at least – by the furious sarcasm of the Reverend Ralph Cudworth, Professor of Hebrew at Cambridge and Master of Christ's College from 1654. Cudworth belonged to an old school of Platonist philosophers who were opposed to Descartes and any kind of empiricism. In a massive work attacking a range of heresies in splendid rhetoric he explained the difference between Epicurean views ('Atomick Atheists') and the *arriviste* hybrid theory of Descartes ('mechanick Theists') that attempted to marry atoms, space and chance to a godly view of creation. And dismissed them both:

God in the mean time standing by as an Idle Spectator of this *Lusus Atomorum*, this sportful dance of Atoms, and of the various results thereof. Nay these *mechanick Theists* have here quite outstripped the *Atomick Atheists* themselves, they being much more extravagant than ever those were. For the professed *Atheists* durst never venture to affirm that this regular *Systeme* of things resulted from the *fortuitous motions* of Atoms at the very first, before they had for a long time together produced many other *inept Combinations*, or *aggregate Forms* of particular things and *nonsensical Systems* of the whole, and they suppose also that the regularity of things in this world would not always continue such neither, but that some time or other Confusion and Disorder will break in again . . . But our *mechanick Theists* will have their Atoms never so much as once to have fumbled in these their motions, nor to have produced any inept System or incongruous forms at all, but from the very first all along to have taken up their places and ranged themselves so orderly, methodically and directly; as that they could not possibly have done it better, had they been directed by the most perfect Wisdom.[34]

Chance and design are like oil and water, or perhaps oil and fire. Cudworth continued more soberly:

There is no *Middle* betwixt these Two; but all things must either spring from a *God*, or *Matter*; Then this is also a *Demonstration* of the Truth of *Theism*, by *Deduction* to *Impossible*: Either there is a God, or else all things are derived from *Dead* and *Senseless Matter*; but this Latter is Impossible; Therefore a God. Nonetheless, that the *Existence of a God*, may be further Directly Proved also from the Same Principle, rightly understood. Nothing out of Nothing *Causally*, or *Nothing Caused by Nothing*, neither *Efficiently* nor Materially.

To which a natural theologian could only add; Amen.

The popularity of the argument from design, and the extraordinary success of Paley's *Natural Theology*, gave wavering Christians a better answer than Cudworth's to the threats of philosophers (deist and atheist) who challenged the basis of Christian beliefs. *By dealing only with existence of God*, without depending on assertions of the authority of God's revelations (in the Bible and in miracles), Paley made an argument for the deist doubter and at the same time created (or at least strengthened) a philosophical context within which contemporary scientists could allay their religious doubts and make a space for their discoveries within orthodoxy. Although not universally admired by those theologians who placed their prime emphasis on revelation, the timeless appeal of the argument from design is shown in the fact that these same threats persist in even more pressing forms today, when our understanding of science has almost limitlessly expanded the realm of Second Causes and a materialist society has put 'belief' and 'faith' onto the defensive.

Francis Bacon had written, in his essay *Of Atheism*: 'A little philosophy makes men atheists: a great deal reconciles them to religion.' By Paley's time, the reverse seemed true. Conventional religious beliefs could be upheld only if one did not probe too far into their philosophical underpinnings. Paley needed to change all that. He knew that he had the gift of reasoning and persuading. And so he set out his proof of God with all the urgency and dedication of a Crusader knight taking arms in defence of Jerusalem. The battleground would have to be all of science and philosophy. In what follows, we must insist on one caveat: it is not fair to judge Paley's *evidence* (or Cudworth's vitriol) by what we know now. It is fair to judge his *conclusions* by such a standard, however, if his arguments are to have any long-standing merit.

CHAPTER FOUR

John Ray: Founding Father

'When you look at a sun-dial or a water clock, you consider
that it tells the time by art and not by chance; how then
can it be consistent to suppose that the world, which
includes both the works of art in question, the craftsmen
who made them, and everything else besides, can be devoid
of purpose and of reason.'

Cicero, *De Natura Deorum*, 77 BC

'If the number of Creation be so exceedingly great, how
great nay immense must needs be the Power and Wisdom
of him who Form'd them all.'
John Ray, *The Wisdom of God Manifested in the Works
of Creation*, 1691

'What absolute Necessity [is there] for just such a Number
of *Species* of *Animals* or *Plants*?'

Samuel Clarke, *Demonstration of the Being and
Attributes of God*, 1705

The central proposition of natural theology is what David
Hume, in *Dialogues*, put in the mouth of Cleanthes (the most
'accurate and philosophical' of his protagonists):

[The world is] nothing but one great machine, subdivided into
an infinite number of lesser machines ... all these various
machines, and even their most minute parts are adjusted to each

other with an accuracy, which ravishes into admiration all men
. . . the curious adapting of means to ends, throughout all nature,
resembles exactly, though it much exceeds, the productions of
human contrivance, of human design, thought, wisdom, and
intelligence . . . By this argument *a posteriori*, and by this argument alone, do we prove at once the existence of a Deity, and
his similarity to human mind and intelligence.

This is the essence of an argument from design and a hundred
years later, Paley's watch analogy said the same thing: 'As for
the watch, so for nature there must exist a Creator.' By extension, the same conclusion must apply to 'every indication of
contrivance, every manifestation of design . . . in the works of
nature; with the difference, on the side of nature, of being
greater and more, and that in a degree which exceeds all computation.' As the watch has a maker, so we have a Maker. As the
watch exists for a purpose, so do we.

When Charles Darwin sat at the window of his rooms at
Christ's College in 1831 reading *Natural Theology*, he found
the arguments 'conclusive . . . the beautiful hinge of a bivalve
shell must have been made by an intelligent being, like the hinge
of a door.' Camped a year later in the Brazilian forest and
seeing at first hand the biological riches of the tropics that
the explorer-naturalist Humboldt had extolled, he wrote in his
journal that 'it is not possible to give an adequate idea of the
higher feelings of wonder, admiration, and devotion which fill
and elevate the mind'.[35] A contemporary anonymous reviewer
of the first edition of Paley's book noted: 'No thinking man,
we conceive, can doubt that there are marks of design in the
universe.'[36] Similarly, in 1876, that quintessentially Victorian
critic Leslie Stephen (father of Virginia Woolf and Vanessa Bell)
praised it – but as if trying not to get his hands dirty: 'The
book, whatever its philosophical shortcomings, is a marvel of

skilled exposition. It states, with admirable clearness and in a most attractive form, the argument which has the greatest popular force and which, duly etherialised, still passes muster with metaphysicians.'[37] In 1996, the biochemist Michael Behe continued the argument seamlessly: 'The reason for the conclusion [that the watch had been designed] is just as Paley implied: the ordering of separate components to accomplish a function beyond that of the individual components.'[38]

The same anonymous reviewer of *Natural Theology* had also grumbled: 'On the subject of Natural Theology no one looks for originality and no one expects to find it.' Given Paley's broadminded approach to borrowing other people's sermons, we should not be surprised to learn that the great watch analogy originated elsewhere and that natural theology itself belonged to a long-standing tradition to which his book simply gave its greatest and most popular expression. Leslie Stephen acidly noted, 'The argument is familiar, and probably has been familiar since the first days when it occurred to anyone to provide a logical basis for theology.' Paley himself called the watch analogy 'not only popular but vulgar' and for contemporary readers it was so familiar an analogy that they would not have thought of attributing the idea exclusively to him. (Fifty years later, enough history had been forgotten that he was accused of plagiarism, the source of these suspicions no doubt lying in the fact that, in accord with the custom of the time, Paley did not supply footnoted references to his sources.) In fact, the watch analogy can be traced back a long way.

In Paley's time, the most immediate exponents of the watch analogy may have been Baron d'Holback (*The System of Nature or, the Laws of the Moral and Physical World*, 1770)[39] or Bernard Nieuwentyt (*The Religious Philosopher, or the Right Use of Contemplating the Works of the Creator*, 1709),[40] who wrote of a man 'cast in a desert or solitary place, where few people are used to pass [coming upon] a Watch shewing the Hours, Minutes and Days of the month'. Hence the charge of plagiarism. Before Nieuwentyt's quite explicit use of the analogy, it occurs in a host of works, including Thomas Burnet's *Sacred Theory of the Earth* (1681), which we shall visit in some detail in a later chapter. Burnet wrote: 'For a thing that consists of a multitude of pieces aptly joyn'd, we cannot but conceive to have had those pieces, at one time or another, put together. 'Twere hard to conceive an eternal Watch, whose pieces were never separate one from another, nor ever in any other form than that of a Watch.' Perhaps the earliest use of the analogy is in Cicero's *De Natura Deorum* (one of the models for Hume's *Dialogues*) where his Stoic philosopher asks: 'Suppose a traveller to carry into Scythia or Britain the orrery recently constructed by our friend Posidonius, which at each revolution reproduces the same motions of the sun, the moon and the five planets . . . would any single native person doubt that the orrery was the work of a rational being?'[41] In fact, as we go along, we will frequently see that several arguments of eighteenth-century scholars consist of little more than a reiteration of what various classical authors had said two millennia before.

One of the great assets of natural theology and the evidence it drew from the world of living animals and plants, is that it was understandable to a broad following who did not have to know code words of contemporary philosophy, or have mastered calculus and chemistry to follow the argument completely. Natural history enjoys a privileged position among the sciences both in its broad accessibility and in the extraordinary aesthetic pleasure inherent in the subject. This is obvious to amateur and professional alike, and only increases the more deeply one probes into the complexities of life. One has only to think of the mechanical perfection underlying the flowing grace of a cheetah in full stride, or the whorled mathematical perfection of a sun-flower. It has therefore always had an extremely wide appeal, whether for a clergyman such as the Reverend Gilbert White who, with his *Natural History and Antiquities of Selborne* (1789), defined the role of the careful observer of local nature in ways that had not been thought about since Virgil and Pliny, or for explorers to the far reaches of the globe like Joseph Banks who travelled with Captain Cook and brought back new natural wonders to test our grasp of the limits of creation.

Above all, nature is always fascinating for what seems to be the common sense, the transparent obviousness, of the fit of its forms to their functions. The elephant's trunk, sometimes powerful, sometimes delicate, is a masterpiece of dexterity far exceeding that of the human hand. The sabre-tooth's canine was a lethal weapon. The barn owl not only has huge forward-directed eyes for detecting its prey, it can also use its super-sensitive ears to pinpoint the source of the slightest rustle of leaves caused by a mouse – in the dark. Some orchids have patterns on their petals that we cannot see, but a wasp, using ultraviolet light and seeing there a picture of another wasp, alights to try to copulate with it and thus unwittingly helps pollinate the flower. Charles Darwin, knowing of a flower with a particular shape, famously predicted that there must exist a kind of moth with a foot-long proboscis to feed deep within it: eventually it was discovered. This is but a tiny set of examples of the exquisite ways in which living organisms are 'adapted' to their environments and 'fitted' to particular ways of life. Such glories of nature have always been the principal evidence that natural theologians adduce for the existence of a creating God – the argument *a posteriori* that Hume allowed as the only possible proof. The vast bulk of writing in natural theology is taken up with elucidating and sermonising upon long lists of such examples from nature; they are the basic evidence for the prosecution's case: such perfections of design and function appear to require us to conclude that a master creator has been at work.

Paley was a logician who lived by the cut and thrust of argument. He added the abstractions of philosophy to the science-versus-religion debate but, as in any great court battle, the case for natural theology was first grounded in hard evidence and that base had long since been constructed by John Ray (1627–1705), its founding father. In many ways Ray and Paley were complementary and opposite. While Paley, at the end of the Age of Reason, depended upon the tightness of his logic, Ray, at the beginning of the scientific revolution, was someone who revelled in facts – both in getting them straight and getting them in order. While Paley was a man who fitted somewhat awkwardly in the machinery of the Church of England, he was nonetheless a true churchman. John Ray was a man of his time – a dissenter.

For every person who is happy to conform, to belong, to submit to the group will, there is always at least one who will not compromise: someone who is sure enough of their own ground to stand apart independently, usually on a matter of principle. It is a great tradition, reinforced periodically by governments who try to force us into what the American poet Emerson called 'a foolish consistency' ('the hobgoblin of feeble minds'). Such was John Ray. He was not just a dissenter; he was a Dissenter. In order to prepare for the Church, he had attended Cambridge in 1644 where his brilliance in science, languages and mathematics quickly showed. He was made a Fellow of Trinity College in 1649, Lecturer in 1651, and sub-Dean in 1658. In 1660 he was ordained as a priest. A stellar career as a Cambridge scholar seemed in prospect, with the living of a rural parish or two to support him and to provide the freedom to pursue his great interest, natural science. Happily

John Ray: Founding Father

John Ray

for us, although he had already published his first book – a compilation of the plants of Cambridge – his timing was bad.

Those were tense years within the state religion and the state itself. The English Civil War was ended, but bitter ill-feelings persisted, particularly among those aristocratic, royalist Cavaliers and their supporters who had lost their lands to the regicide Puritan Roundheads (thus creating a new landed middle class). The broad church that Elizabeth I had carefully nurtured through dozens of compromises had been thrust aside in a passion of radical Puritanism. With the restoration of the monarchy (in the form of Charles II) and election of a reactionary, strongly Cavalier Parliament in 1661, the formal process of retribution began. The obvious target was Puritanism itself and legislation, rather than the sword, was the tool. The Act of Uniformity passed by Parliament in 1662 was not just religious; it was also political, restricting the civil and religious freedoms by codifying the nature of the Church of England and its practices and imposing severe sanctions on dissenters. Instead of a broad church that could tolerate a range of ways of practising Christianity, Parliament opted for conformity. Non-conformists became liable to severe sanctions, including prison or transportation.

All clerics and teachers (and most definitely all fellows of Oxford and Cambridge colleges) were obliged to conform. This meant that they had to worship according to the restored 1549 Book of Common Prayer and swear to the Thirty-nine Articles of 1571 that defined the core doctrines of the Anglican Church. Back in 1643, as a price for their support, Scottish Presbyterians had forced the Roundheads to swear the Solemn League and Covenant which, among other radical measures, abolished bishops and allowed individual congregations to ordain their own priests. The 1662 Act required all clerics of the Church of England to adjure this oath.

Ray had not sworn the oath and had in fact been ordained by a bishop. He was certainly comfortable with the Thirty-nine Articles. But he objected to the coercion; he could not agree that someone who had sworn a sacred oath should be forced to abandon it. At the same time he may already have been restive for greater independence to continue his scientific work. For reasons, particular or principled, that are now unclear, along with 2,000 others he refused to subscribe to the Act of Uniformity.[42] With this, he gave up his fellowship at Cambridge and could no longer teach or preach, although he retained a lay membership of the Anglican Church.

Ray's father was the village blacksmith at Black Notely in Essex, where Ray was born in 1627. Little else is known about his father, but we know that Ray's mother was 'a very religious and good Woman, particularly to her Neighbours that were lame or sick'. Elizabeth Ray was a herbalist healer, which required her to have an excellent working knowledge of botany.

From her, Ray acquired a love of plants, of all nature, of enquiry, and above all an appreciation of the value of precise knowledge. For example, a herbal healer must be able instantly to tell the difference between two very similar looking plants, one edible, the other lethal: the wild parsnip and water hemlock. As Nicolas Culpepper described them in his famous 1652 herbal, *The English Physitian or an astrologo-physiscal Discourse of the vulgar Herbs of this Nation*, the former 'easeth pains and stitches in the sides, and dissolveth wind both in the stomach and bowels'. The latter is 'exceeding cold and dangerous, especially to be taken inwardly'.

If we look at Culpepper or another typical herbal of the period, John Gerard's *Historie of Plants* (1597),[43] with its delightful prose and 1,800 woodcut illustrations, we can guess that John Ray had considerable command of a wide range of plants and their properties, medical and otherwise, even before he entered Cambridge. A lengthy recuperation from an illness in 1650 seems to have given Ray the leisure to explore the countryside and the world of plants more fully. 'First I was fascinated and then absorbed by the rich spectacle of the meadows in spring time; then I was filled with wonder and delight by the marvellous shape, colour and structure of the individual plants.' This soon grew into a systematic study of nature.[44]

Ray's first scholarly book was *Catalogus Plantarum circa Cantabrigiam*, a synopsis of the plants of Cambridge published in 1660 while he was still at Trinity. The obvious next subject would be a botany of all England, an ambition in which he was encouraged by his former pupil and now close friend, Francis Willoughby. In 1662, in his new freedom, he poured himself into this work. His now independent career eventually took him beyond the countryside around Cambridge to further destinations, both geographically and intellectually, than he might otherwise have imagined. Willoughby, no mean naturalist himself and by now not only a friend but a benefactor, proposed a scientific tour of the Continent and in April 1662 Ray, Willoughby and two other Cambridge friends set off on a three-year journey that would take them through France, Belgium, Holland, Switzerland, Austria and Italy. They collected, they discussed science with all the famous men of Europe, and they made notes and drawings of everything they saw and did. This journey set Ray's career firmly on a course that would contribute to changing the religious as well as scientific world.

By 1660, the tightly circumscribed view of the richness and goodness of God's creation as demonstrated by the natural history of Europe had begun to be overshadowed by the abundance of plants and animals brought in from the rest of the world by explorers and merchants. For Ray and his contemporaries across Europe, exploration in every corner of the globe had opened a Pandora's box of nature. If this was God's creation, suddenly it had unfathomable, incomprehensible depths of diversity. Noah's ark could not have held a thousandth part of the living animals and plants with which natural philosophers were confronted, as travellers brought back to Europe every kind of unimagined creature – some real but improbable (like the kangaroo), others (like the mermaids and unicorns) fabulous yet all too believable. In the last half of the seventeenth century it was relatively easy to know at least eighty per cent of the plants of Britain and difficult but not impossible to know sixty per cent of the plants of western Europe. Ray thought that there were some 10,000 kinds of insects, 1,300 other kinds of animals and 20,000 species of plants in the world. By 1750, those estimates were debatable; by 1850 they were laughable. Today the worldwide total of known living species is 1.8 million and rising.

All of nature, whether an English woodland, the patch of Brazilian rainforest where Darwin sat to write his journal, or the biological diversity of a whole continent, is an apparently chaotic arrangement of organisms: flies next to birds next to worms, in trees, over grasses. In terms of ecological interactions – a kind of pattern or machine – they can be shown to make perfect sense. They constitute networks of spatial and temporal relationships behind which lie intertwining chains of cause and effect. With the discovery of order comes the possibility of finding lawfulness. For example, certain kinds of flowers (primroses, daffodils) grow in woodland and they 'must' flower before the leaves of the canopy emerge. Such flowers are almost invariably yellow and white, never red. Red flowers occur where levels of incident light are higher and they usually bloom in the open, later. Similarly, if one looks at physical structure (anatomy), which is what anyone sees first, an order appears. Of the birds in Britain or Europe or the world, all the woodpeckers fall together, separately from all the finches, the ducks and so on. It was perfectly obvious to every countryman that Queen Anne's Lace, cow-parsley, wild parsnip and hemlock were all similar to each other, and different from celandines (which in turn seem closer to buttercups). If this is lawful, we can speculate about the causes.

Before anyone could make any sense out of the confused and confusing mass of new information about nature, the data had to be organised; what, for example, a woodpecker or a celandine *is* had to be defined. Let us imagine, for a moment, the contents of an automobile-parts warehouse dumped, thousands of different bits and pieces. You have to store them and then use them; the first steps must be to organise everything and put things on shelves where they can readily be found: all the brake pads in one place, the light bulbs in another. But cables would present a problem; whether to put brake cables with the other brake parts or with, say, the accelerator cables. There is probably no one perfect way to organise car parts.

Imagine, then, doing all this without any notion of what a

car is. You couldn't do anything logically unless you knew what the bits were for. You couldn't discover what the bits were for until you had at least understood them enough to categorise them: a circle of ignorance. This was the problem for the natural philosophers who tried to categorise nature in order to understand it better and, not incidentally, to discover what it told them about the Creator. They had to create systems of classification that discovered patterns out of what would otherwise be a random arrangement of entities. For example, given a disassembled car one would quite easily work out from the external body panels that automobiles are bilaterally symmetrical, with a front and back, left and right. But then the five, rather than four, wheels with road tires would make no sense – that would require a new concept, that of the spare wheel. And the steering wheel would be baffling; one's natural inclination to classify it with the other wheels would likely slow or even prevent any attempt to discover a different use for it. In the same way, when Aristotle made his first classifications of living things, he separated the whales from the other mammals: logical enough in a way. But he also correctly saw that whales were quite different from fishes, even though both lived in the sea. In any case, we can excuse Aristotle his errors, remembering the wise man who once said 'never do anything for the first time'.

Ray's passion to organise the facts of nature was not mere stamp collecting but a search for new, deeper truths. The patterns (classifications) that Ray and those following him established were philosophically powerful. A surprising depth of enquiry into the root causes of things is bound up in the apparently simple statement that the flightless dodo is related to the pigeon rather than the ostrich. Once one has found patterns in the distribution of different kinds of organisms in the world, one is naturally led to queries about the causes of those patterns. If God made them, then the patterns are a reflection of the symmetry and orderliness of God's perfect mind. The sub-patterns might then be the result of God working out variations of different ideas – the idea of a worm, a woodpecker or a pigeon, for example. But this also depended on one of Ray's greatest contributions, which was to establish the building bricks, or the least common denominators, out of which the great natural groupings of organisms are made. The early practice and theory of classification honoured the biblical practice of referring to distinct 'kinds' of creatures. We now call such entities 'species' and we are so used to the idea that each species of animal and plant is different from all others, and has its own name by which we distinguish it from its sisters, that we simply take for granted the concept of species itself, and even that names might be important. This concept of the species ('species' is the Greek for 'kind') was John Ray's lasting contribution to natural science.

An instructive by-product of the identification of these patterns comes in the discovery of the very opposite. Because classifications aim to produce rational patterns, they prompt the investigator to query the apparently irrational that is also revealed. For example, many of the 'natural' groupings of animals and plants appear to have a geographical cause or at least a geographical consistency. This leads to new questions: Why, for example, are there no penguins in the Arctic, or polar bears in the Antarctic? One could understand why God might have created parrots for the tropics but why are hummingbirds only in the New World? Even more puzzling is the fact that Europe and North America each have eagles, kingfishers and woodpeckers – but different kinds. These inconsistencies (which dogged Darwin in his Cambridge and *Beagle* days and helped lead him to ideas of evolution) hinted of chance rather than purpose and any time the word 'chance' cropped up in 1800 (or 1700) it brought with it the possibility that the pattern was the result of contingencies in the underlying process, and particularly it raised the spectre of the 'chance collisions of atoms' and all the other Epicurean and Cartesian Second Causes.

Another difficulty slowly to emerge was that, with the identification of the species as the basic entity in classification, a circular argument had developed. On the one hand, one can only list and systematise discrete and fixed entities. One cannot systematise the constantly changing parts of a cloud or the molecules of steam escaping from a boiling kettle. On the other hand, the very fact of naming and classifying species established and reinforced their fixity. Immutability of species became both a necessary premise and a *consequence* of the science of classification. The very tools that opened up the world of biological diversity (by making it rational) tended to close off any discussions of its basis in fact (by denying the possibility of change). And of course this was a boon to theology. If they were created by the Almighty, the species of animals and plants on earth must today retain the individuality with which he originally endowed them. If God at creation 'found them good', that allowed (if it did not dictate) the logical conclusion that he made them fixed and immutable, unchanging and unchangeable. From then on, there was huge pressure to see species as real and immutable, despite all evidence to the contrary.

Those who argued the other case, as did fledgling evolutionists, a hundred years later were more easily dismissed. Even without attempts to interpret Genesis to that effect, the whole concept of God having created living organisms implied that they were perfect – a reflection of the perfection of God's mind – and would not change. It must, therefore, have been more than a little disconcerting to the great eighteenth-century botanist

Linnaeus and his contemporaries eventually to discover that species were not fixed after all. Indeed, the idea of fixed species flew in the face of common-sense experience with cross-breeding, both artificial and in nature. It was not consistent with the notion of discrete entities that they should be able to interbreed and produce what looked like new kinds of animals and plants. Hybridisation therefore became a worrying problem, both among the new breed of systematists who, like Linnaeus, could look to it as a mechanism by which at least some of the different species might have arisen, and also for those who believed that God had created each species perfectly, fixed for ever in time and space.

Throughout his extremely productive career as a natural scientist, scratching out a living from writing study after scholarly study elucidating the patterns of nature, John Ray remained all the while faithful and true to his religious beliefs, all the time devoted to the central premise that nature was the handiwork of God and that species were immutable. In thirty years of studies and travels he had revolutionised 'natural science' (botany and zoology), almost single-handedly moving it from a medieval to modern mode. And, while his religious life had been curtailed by his formal separation from the Church, his instincts as a teacher and preacher remained. Relatively late in his career he turned back to write out his philosophy as a Christian and a scientist. *The Wisdom of God Manifested in the Works of the Creation*,[45] based on lectures and sermons he had given around 1660 when still at Cambridge, became the first English manifesto of natural theology and his most popular and influential work.

For Ray, as for St Thomas Aquinas, the idea that the principal independent evidence for the existence of God was the diversity and perfection of his creation simply represented common sense:

You may hear illiterate persons of the lowest Rank of the Commonalty affirming, that they need no Proof of the being of God, for that every Pile of Grass, or Ear of Corn, sufficiently proves that. For, say they, all of the men of the World cannot make such a Thing as one of these; and if they cannot do it, who can, or did make it but God? To tell them that it made it self, or sprung up by chance, would be as ridiculous as to tell the greatest Philosopher so.

There is no greater, at least no more palpable and convincing Argument of the Existence of a Deity than the admirable Art and Wisdom that discovers itself in the mode and constitution, the order and disposition, the ends and uses of all the parts and members of this stately fabrick of Heaven and Earth. For if in the world of Arts, as for example: a curious Edifice or machine, counsel, design, and direction to an end appearing in the whole frame and in all the several pieces of it, do necessarily infer the being and operation of some intelligent Architect or Engineer, why shall not also in the Works of nature, that Grandeur and magnificence, that excellent contrivance for Beauty, Order, use, &c. which is observable in them, wherein they do as much transcend the Efforts of human Art and infinite Power and Wisdom exceeds finite, infer the existence and efficiency of an Omnipotent and All-wise Creator?[46]

In other words, as the Psalmist put it, 'The Heavens declare the Glory of the Lord and the Firmament sheweth his handy work.' (Psalms 19:11) In the Age of Reason and Enlightenment it was entirely natural, to churchmen at least, that the application of all man's powers of observation and logic, applied to nature, could only extend and confirm our appreciation of the powers of God. A close study of nature would reveal, according to the natural theologians, such an intricacy of construction and such a close fit of form to function, that nature could only have arisen through the active power of some great creative force. God was seen as the supreme celestial architect and mechanic. The glories of the natural world – the grace of a panther, the symmetry of a spider's web, the heady scent of a rose – all made the case for a benevolent Designer.

This, then, was natural theology. Surely only a wise and all-seeing God could have created all the living creatures, from the tiniest ants to the majestic elephant and the slightly absurd giraffe. These natural wonders could not have arisen through mechanistic chance, and the great point of reconciliation between natural science and religion was therefore that natural science, through local observation and world-wide exploration, was steadily revealing wonder upon new wonder to confirm the scope of God's handiwork. Science and religion were not opposed after all.

On the other side 'of the debate', Ray's contemporary, Robert Boyle, had espoused a watered-down version of Descartes' neo-Epicurean view of a world driven by the chance collisions of atoms in empty space, in which order results from the lawfulness of matter itself. At one point Boyle had written:

> I think it probable, that the great and wise Author of things did, when he first formed the Universal and Undistinguished Matter into the World, put its parts into different Motions, whereby they were necessarily divided into numberless Portions of different Bulks, Figures and Situations in respect of each other. And that by his Infinite Wisdom and power he did so guide and over-rule the motions of these parts, at the beginning of things as that (whether in a shorter or longer time reason cannot determine) they were finally disposed into this Beautiful and orderly Frame, that we call the World; among whose Parts some were so curiously contrived, as to be fit to become the Seeds or seminal Principles of Plants and Animals. And I further conceive, that

he setled such laws or Rules of local Motion, among the parts
of the Universal Matter, that by his ordinary and preserving
Concurse the sevral parts of the Universe thus once completed,
should be able to maintain the great Construction or System and
Oeconomy of the mundane Bodies, and propagate the Species
of all living Creatures.[47]

Such an account of creation has a great deal of Newton's clockwork universe in it. Ray instead propounded the central element
of natural theology, that the complex structures of the natural
world could only have been created under the influence of a
Supreme Intelligence, according to His plan, and for His purposes. 'Intelligence seems to me requisite to execute the Laws
of Motion . . . I am difficult to believe that the Bodies of Animals
can be formed by Matter divided and moved by what Laws
you will or can imagine, without the immediate Presidency,
Direction and Regulation of some Intelligent Being.' Here was
where Ray's certainty – certainty of his maker and certainty of
his encyclopaedic knowledge of the natural world – came to
the fore. He was able to overwhelm his opponents, and bolster
the doubters, with facts.

As Ray expounded it, in natural theology there is a twofold
demonstration of God's creative hand. The sheer complexity of
nature makes the case for the first half of the argument from
design – all the complexity of structure and function that Ray
painstakingly recorded and classified inevitably led to the conclusion that they could only have been caused by an intelligent
creator. The second half of the argument is that the same complexity of design in nature is 'good'. Its perfection and purpose
(its fitting for use) must be the result of a perfect, purposeful
intelligence and is specially manifest in what we call adaptation:
the matching of form and function. And adaptation is something that can be documented (or, more scientifically, it is a

proposition that can be tested) by empirical observation and even experiment: the leg of the horse is 'obviously' adapted for running fast; the leg of a deer is superficially quite similar, presumably for the same purpose, but achieves this with two elongated toes where the horse has but one; the foot and foreleg of a mole are quite different, being broad, short and stout, ideal for digging; the duck's webbed foot is obviously suited to swimming; the long delicate toes of the jacana allow the bird to walk over waterlily pads. In each case the structure of the leg or foot fits with a particular use. Similarly, the heart is perfect for pumping blood, the kidneys for filtering it, the lungs for charging it with oxygen: that is adaptation.

A great deal can be made of this concept, whether in the sense of a mechanism that performs a specific function (like a watch or the human heart), or in the ways that the constituent parts work together (as in the superbly dextrous human hand), or in the sense of a perfect fit between organism and context (the lock-and-key adaptation of the beak of a particular species of hummingbird to the shape of the flowers on which it feeds and which, in the process, it fertilises). Ray's strategy in *Wisdom of God*, and the strategy of natural theologians ever since, was to overwhelm the reader with so many examples of this kind of adaptation, and of such complexity, as to make it impossible to imagine that chance or anything but mediated, purposeful design had been their cause. The list of examples that might be used is virtually endless. Ray attempted a synopsis of all relevant science, from cosmogony, fossils, air and water, the 'uses of flowers', and the adaptation of form and function in animals, concluding with a detailed discussion of human morphology and physiology. Everything was grist to his mill, from the motions of the planets to the migration of birds and the causes of rain.[48] All together, *Wisdom of God* gives a fascinating snapshot of the state of natural science at the end of the seventeenth century – a science on the cusp of modernity.

Paley, more than a hundred years later, had the advantage of knowing more about physiology; in his *Natural Theology* he founded his case on examples from the internal workings of animals, particularly humans. He included whole chapters on the mechanics of the human skeleton, muscle systems, blood vessels, animal physiology and comparative anatomy, and several on special adaptations, including instinct, insects and the structure and function of plants. His *Natural Theology* is almost a textbook of comparative anatomy and physiology. There is scarcely any point, he feels, in elaborating on these lists, these catalogues of organisms showing both genuine complexity of structure and an apparently close fit between form and function. Nature as a whole *appears* to show a general fitted-ness and purposefulness – adaptation – that extends to whole ecosystems. Discovering the intricacies of adaptation and the orderly interconnectedness of nature was an ideal pursuit in the Age of Reason.

Unfortunately, side by side with all the rationality of the Enlightenment, there existed also an Age of Unreason in the form of a naive utilitarianism. Ray had little or no hesitation in affirming that not only had God never made anything in vain, he had made all (or most) things for man's benefit. As Paley later put it: 'How close is the suitableness of the earth and sea to their several inhabitants . . . and of those inhabitants to the places of their appointed residence.' He evidently had no qualms for example, in finding an example of God's goodness to man in the relationship of 'sleep to night', stating that God had provided the separation of night from day (Genesis): 'Were the bustle, the labour, the motion of life, upheld by the constant presence of light, sleep could not be enjoyed . . . it is happy therefore for this part of creation . . . that nature . . . has commanded, as it were, and imposed upon them, at moderate intervals, a general intermission of their toils, their occupations, pursuites.' It is only reasonable to prefer this conclusion to its converse – that the human diurnal cycle is an adaptation to a cosmological necessity – if one is sure that the world has been created *for* man. (To his credit, Paley was careful not to insist that everything in creation was beneficial, but only the 'vast plurality . . . of contrivance'.[49])

Utilitarianism is one of those superficially attractive philosophical positions that self-destructs under closer scrutiny. It might be plausible that God made, for example, apples sweet and colourful so that men would know they are useful food to eat. But why is the sea salt? A utilitarian answer is that the salt in the sea is useful to savour our food. Similarly, the tides were made to float ships in and out of harbour; rocks exist for us to build our houses; horses have been created at just the right height for man to ride. The bountiful hand of God can be seen as acting everywhere in nature and this – because it is pious and religiously sound – seduces us into closing our eyes to the facts. We forget to look for other explanations. An independent viewpoint cannot ignore the fact that the sea's salt is a curse, because it means that ninety-five per cent of the earth's water is not drinkable – a fact known only too well to any shipwrecked sailor; for half of every day, the tides are against us; most rocks are far too heavy and hard, or alternatively too soft, readily to be useful for building; dogs are too short to ride; horses are hopeless at climbing trees or burrowing after rats. Furthermore, all of nature is functionally interconnected, not human-directed. Elephants leave in their wake vast quantities of dung that, in turn, is food for countless dung beetles. The larvae of dung beetles in their turn are food for birds. From a utilitarian human viewpoint this is a satisfyingly tidy system; the beetles clean up after the elephants and the birds keep the beetles in check. For a dung beetle, however, the more dung the better and the elephant is merely its necessary source. For the bird, the elephant is the ultimate provider of large numbers of beetles.

Of all the examples that have been produced to prove the impossibility of nature having been anything but the product of God's designing, purposeful intelligence, none has been more popular or convincing than the human eye: 'Examination of the eye [is] a cure for atheism,' as William Paley said. While each of the various writers on natural theology from 1700 to the present has had his favourite examples of the works of the Almighty, all of them have returned eventually to this. The eye is a wonderful piece of adaptation of form to function. It appears to be exactly like a machine of human invention. And self-evidently, the example of the eye ought to destroy any rival theories of change and evolution – for what good would a half-evolved, half-formed eye be? Book after book produced the eye as its key evidence. Book after book ridiculed the notion that the eye could have been produced by some material process of change, especially by the 'chance motion' of atoms or corpuscles out of which atheists like Epicurus, Descartes and Buffon thought the complex objects of this world had been assembled.

The eye seems such a good example because it so perfectly mimics a 'machine' of human devising. For us moderns it looks like a camera, whereas in fact the camera mimics an eye. The pupil that controls the amount of light entering the eye acts just like the diaphragm that changes the f-stop on a camera. There is a lens that can focus an image on a receiving structure, the retina. The retina consists of light-sensitive cells connected by nerves to the brain. Looking at the light-receptive cells, we can imagine each one to be a tiny digital camera with its cable (nerve fibre) snaking off to the central processing unit (the brain).

Before the camera, the comparison was with a telescope: 'I know of no better method of introducing so large a subject, than of comparing ... an eye ... with a telescope. As far as the execution of the instrument goes, there is precisely the same proof that the eye was made for vision, as there is that the telescope was made for assisting it.'

The telescope and the microscope are intimately related. When Galileo turned his telescope to the heavens, his very first nights of observation set in chain a revolution. Miraculously, a telescope turned upside down – a microscope – reveals yet another new and different world. Various versions of an elementary microscope existed from the mid-sixteenth century. Today no school biology class would be complete without the *sine qua non* of microscopy – an amoeba oozing its way from one shape to another, exactly as the books describe. Except that no book could capture the miracle of this blob of jelly constantly reinventing itself into another shape, constantly on the move and yet going nowhere, all the while its innards in some kind of slow rolling boil, like a semi-transparent porridge being cooked in very slow motion. Was this how we all began, billions of years ago? No science fiction was ever as fascinating as a real amoeba. But when we look down a microscope, we have some prior inkling of what to expect. It is almost impossible to imagine the wonderment of the first men and women to witness such miracles, men like Robert Hooke (naturally, he was at the forefront of microscopic technology too[50]) and the Dutchman Antoni Leewuenhoeck, looking for the first time ever at a bee's sting, a fly's compound eye, a louse or a flea, and discovering a wealth of detail, an exquisite array of biological engineering to gladden any natural theologian's heart.

When Galileo turned his telescope around and looked at the minute details of nature, instead of the vast empty spaces of the universe he found a world that offered more support to the believer, at least at first. The microscope was in many ways the best possible of tools for the natural theologians. Within the world of contrivance seen with the naked eye lay further realms of even more intricate complexity and beauty. The microscope opened up an inner world of perfection in miniature, of intricate and inspirational detail, and of unimagined variety. Where observers had previously seen complexity of structure and precision of adaptation, the microscope showed that this existed on such a grand scale – a smaller physical scale but a greater degree of complexity – that they recorded it with the same awe and sense of devotion as did Darwin his rainforest. Hooke's great book, *Micrographia* (1665), was the first treatise on miscroscopy, and also a vehicle whereby he poured out his ideas on a wide range of subjects, from optics to earthquakes. He described the wonderful details to be seen in a common flea:

> But, as for the beauty of it, the Microscope manifests it to be all over adorn'd with a curiously polish'd suit of sable armour, neatly joint'd, and beset with multitudes of sharp pinns, shap'd almost like a Porcupine's Quills, or bright conical Steel-bodkins; the head is on either side beautify'd with a quick and round black eye, behind each of which also appears a small cavity, in which he seems to move to and fro a certain thin film beset with many small transparent hairs.[51]

Looking down a microscope at the head of a flea was just like opening the back of a beautifully made watch and seeing for the first time all the glowing, moving parts that make it work. But no watch of Paley's imagining, however complex in its motions, could match the wonders revealed by just one hour at a microscope.

CHAPTER FIVE

Difficulties with the Theory, and the Argument Extended

'All things bright and beautiful,
All creatures great and small,
All things wise and wonderful,
The Lord God made them all.'

Mrs C. F. Alexander,
'All Things Bright and Beautiful', 1848

'I am difficult to believe, that the Bodies of Animals can be formed by Matter divided and moved by what Laws you will or can imagine, without the immediate Presidency, Direction and Regulation of some Intelligent Being.'

Robert Boyle, *A Free Enquiry into the vulgarly receiv'd Notion of Nature*, 1685

'How can the events in space and time which take place within the spatial boundary of a living organism be accounted for by physics and chemistry?'

Erwin Schrodinger, *What is life?*, 1944

John Ray had thrust the biological sciences into the centre of theological debate, and they would remain the foundation on which to build a natural theology. With new wonders of the natural world revealed every day, the argument from design seemed robust. In truth, however, by the end of the eighteenth century biology had proceeded very little beyond description and classification, and old ideas were just as hard to shed there as anywhere in the human experience. A great deal was passed along as fact when it was only the repetition of myth and superstition. While experiment was well established as the core of the physical sciences such as chemistry and physics (the contemporary name for which was 'Experimental Philosophy'), in the biological sciences experimentation was still in its dark ages. (Not until 1786, for example, did Galvani discover the electrical basis of the nerve impulse.) The study of living organisms was largely observational and even then partially held back by the weight of classical authority. In 1788, the Reverend Gilbert White wrote his *Natural History and Antiquities of Selborne* which has become one of the best-loved examples of writing on nature; but it reveals a state of exquisite uncertainty about scientific issues such as the behaviour of birds. Intellectually, like so many of the authors of his day, White (1720–1793) was a man on the cusp between the old descriptive methods of Virgil and the modern Newtonian era.

The young White suggests a little of what Charles Darwin might have become. He graduated from Oxford having taken sport at least as seriously as learning, and had drifted into a career in the Church. A fellowship at Oriel College gave him just enough money to follow his own interests and he never progressed in the Church beyond a number of curacies – looking after parishes for richer, absentee clerics. His dual passions were the Hampshire countryside around his beloved Selborne (where his grandfather had been vicar, a position to which he could never aspire as the living was in the gift of Magdalen College) and natural history. Like so many of his age, he was an inveterate note-taker and diary keeper and he found early inspiration in Philip Miller's *Gardener's Dictionary* (1739) and John Ray's *Methodus Plantarum Nova* (1682).

But White was not interested in the minutiae of identifications and classifications. He disapproved of 'faunists', who were 'too apt to acquiesce in bare descriptions, and a few synonyms; the reason is plain; because all that may be done at home in a man's study; but the investigation of the life and conversation of animals is a concern of much more trouble and difficulty, and is not to be attained but by the active and inquisitive, and by those that reside much in the country'. Over the years of walking and working in Hampshire, the long horseback rides in all seasons to the various churches under his care (carriages made him sick), and his daily labours in the garden at the Wakes, the house in which he was born and died, he became something of an expert naturalist. In 1751 he began a formal 'Garden Kalendar' in which he kept notes on the weather, times of planting and harvesting, first sightings of birds, and records of their songs and their habits. He might have remained in total obscurity except that in 1761 the naturalist Thomas Pennant began to write a four-volume synoptic account of *British Zoology* for which the publisher was White's brother Benjamin in London. Gilbert White discovered many errors in Pennant's work and the two corresponded. Soon a third enthusiast was added to the correspondence, the Welsh judge Daines Barrington. White's diaries and his letters to Barrington form the bulk of the *Selborne* book.

White was at the leading edge of contemporary observational natural history. And therein lay the problem – a conflict between what one could see for one's own eyes and what the accepted authorities (in his case, the classical authors of antiquity and the Middle Ages) had said. Perhaps the most glaring example of this concerned bird migration. White delighted in logging the appearances of wandering birds normally rare to England, such as the black-winged stilt and hoopoe – more familiar to Pliny and Italy than to Hampshire farmers. He wrote to Pennant, in a letter that captures his easy conversational style: 'The most unusual birds I ever observed in these parts were a pair of hoopoes . . . which came several years ago in the summer, and frequented an ornamental piece of ground, which joins to my garden, for some weeks. They used to march about in stately manner, feeding in the walks, many times in the day; and seemed disposed to breed in my outlet; but were frightened and per-secuted by idle boys.'

White very much wanted to believe the evidence that birds like swallows and swifts migrated to and from Britain annually. Another brother, something of a black sheep who had been packed off as chaplain to Gibraltar, reported to him from the colony that he had seen 'vast migrations not only of Hirudines . . . but also of many of our soft-billed birds of passage' flying overhead towards Britain in the spring. Weeks later, they would arrive. But something in White could not give up the medieval idea that 'many of the swallow kind do not depart from this island, but lay themselves up in holes and caverns, and do, insect-like and bat-like come forth at mild times, and then retire again to their . . . lurking places'.

At the same time, White adopted a more modern experi-mental approach to answer the question of why cuckoos lay their eggs in other birds' nests. One theory was that cuckoos lacked the necessary body structure to brood their eggs. White tackled the question by dissecting a cuckoo and its close relative the nightjar, the latter being a bird that brooded its eggs nor-mally, to look for differences that would explain their be-haviour. The two were identical. Therefore, the reason for brood parasitism must lie somewhere else: 'We are still at a loss for the cause of that strange and singular peculiarity in the instance of *Cuculus canorus*.' White was also modern enough to be actively involved in early experiments with smallpox inoculation in Selborne, tested whether bees could hear by the simple expedient of shouting at his hives with a speaking trumpet (they ignored him), and even tried to measure the speed of sound using echoes.

Interestingly, despite his life as a naturalist-cleric, one does not find in White any reference to John Ray's brand of natural theology. This was perhaps because White's world was one of observation and classical authors, rather than philosophy. It was a harsh world that would have recognised the Victorian hymn quoted at the head of this chapter but found it overly sentimental. In religious terms White was wholly orthodox, belonging to the world in which personal faith in God was more important than so-called proofs from science. 'All Things Bright and Beautiful', however, perfectly sums up the natural theologian's temptation to ascribe all the diversity and wonder of nature to the hand of a wise and wonderful God, even though they would rather carefully ignore the second half of the biblical source of that poem: 'He hath made every thing beautiful in its time: also he hath set the world in their heart, *so that no man can find out the work that God maketh* from the beginning to the end' (Ecclesiastes 3:11, emphasis added).

Even if we grant, with Ray and Paley, the apparently precise fitted-ness of complex living structures to unique functions and particular circumstances, 'easily suiting the Fabrick of the parts to the Uses that were to be made of them;'[52] we still have to return to the stubborn central question, the fundamental challenge: to move beyond the watch analogy to direct evidence that proves this is all the work of an intelligent contriver. Ray was not an experimental scientist; he worked by collecting examples from nature. Paley tried to find a second order of truth by synthesising these observations into generalisations that went beyond individual circumstances. 'Symmetry' seemed such a case to Paley: 'the exact correspondency of the two sides of the same animal . . . with a precision, to imitate which in any tolerable degree forms one of the difficulties of statuary, and requires, on the part of the artist, a constant attention to this property of his work, distinct from any other.' And, without a hint of irony, Paley gave us also 'asymmetry', the component of design allowing various parts to be packed together in a small space and still function efficiently. For example, in the human abdomen, 'the contents have no such correspondency [symmetry] . . . the heart lies on the left side, the liver on the right side.'

A third generality shows how readily one can conflate material and immaterial properties and it got Paley into yet another metaphysical argument that he could not resolve. For it is 'Beauty . . . not relative beauty, nor that of some individual compared with another [but] generally the provision which is made, in the body of almost every animal, to adapt its appearance to the perception of the animals with which it converses.' Here Paley defined beauty as the 'wonder that natural objects inspire' and then went straight on to ask whether there is 'such a thing as beauty at all'. Rather, he wondered whether 'whatever is useful and familiar comes of course to be thought beautiful; and that things appear to be so, only by their allegiance to these qualities . . . our idea of beauty is capable of being so modified by habit, fashion, by the experience of advantage or pleasure; and by associations arising out of that price, that a question has been made, whether it be not altogether generated by these causes . . . I would rather argue this.'

His fourth general property of nature was 'relation' – the interacting nature of the various parts of an organism – and here he was on more familiar ground, for no one could find a better analogy for the fitting together of the muscles, bone, sinews, blood vessels and nerves of an organ like the arm than 'the several parts of the watch, the spring, the barrel, the chain, the fusee . . .' In this sense, 'relation' is the essence of the adaptation of the organism to its function and its environment. A final generality bespeaking an intelligent creator is 'compensation', which is actually the flip side of relation: 'when the defects of one part, or of one organ are supplied by the structure of another part, or of another organ ... the short unbending neck of the elephant is compensated by the length and flexibility of his proboscis' and so on. All this represents a category of adaptation such that 'the bodies of animals ... hold a strict relation to the elements by which they are surrounded'.

The Reverend Samuel Clarke (1675–1729) had a rather clever contrarian view of things that would have been useful to Paley if it had not also contradicted the latter's 'generalities' of adaptation. Clarke was yet another of the series of gifted preachers and scholars to come out of Cambridge at the turn of the eighteenth century, all heavily influenced in one way or another by the great Newton. Clarke was a brilliant but unorthodox metaphysician who was damned on the one side for flirting too closely with rationalist deism and the Arian heresy and on the other for not freeing himself from the constraints of faith and belief. A follower of Newton and Leibnitz, he was adamantly opposed to John Locke's sceptical philosophy and was invited to give the first two series of Boyle Lectures (in 1703 and 1704), endowed by Boyle himself to promote natural theology.[53] With respect to the perfection and consistency that most natural theologians found in nature, Clarke argued: 'all Things in this World appear plainly to be the most arbitrary that can be imagined; and to be wholly the Efforts, not of necessity, but of Wisdom and Choice. A Necessity indeed of Fitness; that is, that things could not have been otherwise than they are, without diminishing the Beauty, Order, and well-being of the whole.'[54] In other words, the evidence for God is not pattern and symmetry in nature but the very *disorderliness* of it all. Where other natural theologians were forced by their devotion to see a perfect and symmetrically ordered world, Clarke was content to see the true hand of God demonstrated in his having made the kinds of choices that blindly acting nature never could.

David Hume, however, had already made this a question far too serious and complex to be solved through Clarke's kind of debating point. In the *Dialogues*, Cleanthes points out that 'Two eyes, two ears are not absolutely necessary for the subsistence of the species. The human race might have been propagated and preserved, without horses, dogs, sheep, and those innumerable fruits and products which serve to our satisfaction and enjoyment.' Cleanthes means this as an argument for a 'benevolent design' but we can turn it around and ask: why not a man with one eye and two noses? The modern answer lies not in the perfection (or otherwise) of a particular design, but the contingencies of history.

Hume's Philo insisted that adaptation was too easy a human construct. 'It is in vain . . . to insist upon the uses of the parts in animals or in vegetables and their curious adjustment to each other. I would fain know how an animal could subsist, unless its parts were so adjusted.' In this case, adaptation is not the result of some purposive end-directed process, but simply the result of the winnowing out by nature of all the combinations that don't work, leaving those that do. This produces the 'appearance of art and contrivance . . . all the parts of watch must have a relation to each other and to the whole . . . a defect in any of these particulars destroys the form; and the matter, of which it is composed, is again set loose, and is thrown into irregular motions and fermentations, till it unite itself to some other regular form.' Hume had no firm idea of what he actually meant by this; the whole argument is set in a vague atomistic mode. But we can read it, more than 200 years later and with the rose-tinted spectacles of hindsight, and see the presaging of a distinctly Darwinian natural selection.

Difficulties with the Theory, and the Argument Extended

If Hume was right, then there was a new answer to the next question (framed first by Lucretius): whether it was true, that in any apparent 'adaptation' the structure was 'caused' in order for a particular use or, instead, whether the particular use arose out of the (prior) nature of the organ.[55] For example, in the case of the long, curved bill of a hummingbird, that shape of bill might have been caused (whether by chance or by God) in order to fit a particular flower, or the bill might have been caused first and then turn out (only) to be useful in feeding from that kind of flower. The answer might be that, out of the various combinations that nature constantly throws up, by chance some turn out to work, most don't. Clarke's ideas can therefore be seen as an attempt to counter the atheist/deist notion that nature was simply the end product of 'what works'. William Paley tried a different counter to Hume:

> So far as this solution is attempted to be applied to those parts of animals, the action of which does not depend on the will of the animal, it is fraught with still more evident absurdity ... faculties thrown down upon animals at random, and without reference to the objects amidst which they are placed, would not produce to them the services which we see: and if there be that reference, then there is intention.

In other words, he simply asserted that there is no experimentation in nature and there were no errors at creation. This rejection of Hume's perfectly logical propositions makes sense within Paley's narrow framework. Adaptation could not be recast as an element in the system of Second Causes that we call evolutionary mechanism by means of any modest adjustment in that view. That would depend upon a wholesale revision, based on an articulation of a possible material cause or mechanism (natural selection, for example). Paley's world view was confined to fixed

91

entities, each matched perfectly by an independent, external being to their functional setting. Hume's view was of change. Similarly, Darwin's theory produces a world forever changing over geological time, in which new variants of organisms are constantly produced and equally constantly tested against a pre-existing environmental context, some succeeding, some failing. Paley's is a view of matter in stasis; Hume's and Darwin's is one of constant motion. Paley's has the certainty and the independence from natural laws of the First and Final Causes; Darwin has all the chances and uncertainties, and all the lawfulness, of Second Causes.

One of Hume's most forceful arguments against natural theology was to ask whether the whole endeavour did not simply beg the question. Natural theology has a validity only as long as one is able to see the works of nature as perfect. While Clarke thought the irregularities of nature attested to a mind rather than inviolable laws, Hume offered the contrary view: nature is fundamentally imperfect and no one can discover a great transcendent intelligence in 'the inaccurate workmanship of all the springs and principles of the great machine of nature ... one would imagine that this grand production had not received the last hand of the maker; so little finished is every part, and so coarse are the strokes with which it is executed'. Hume's challenge, that Paley had to answer, was to say that, in essence, the appearance of design in nature is simply our rationalisation of 'what works', and its supposed perfection is functionally 'that which is good enough'. That was certainly the consequence of Charles Darwin's evolutionary theory of natural selection, according to which, in a changing world, everything must change in order to keep up.[56] While the popular complaint has always been that an 'adaptation' half-formed would be useless (a giraffe with a medium-length neck or an elephant whose trunk did not reach the ground), Darwin's theory postulates that at all stages of change, the new works better than the old in the new circumstances.

With this, we have to return to one of the keystones of design theory – the eye. Charles Darwin knew that explaining the eye was essential to the success of his theory. In the sixth chapter of *On the Origin of Species* ('Difficulties on Theory') he asked the reader to consider how one can trace a gradation of different types of 'eye' and demonstrated how easily it could have arisen through gradual change over millions of years. 'Reason tells me, that if numerous gradations from a perfect and complex eye to one very imperfect and simple, each grade being useful to its possessor, can be shown to exist . . . and if any variation or modification in the organ be ever useful to an animal under changing conditions of life, then the difficulty of believing that a perfect and complex eye could be formed by natural selection . . . can hardly be considered real.' As knowledge of the light-sensitive organs of simple creatures has increased in the last 150 years, Darwin's explanation has only gained in strength; the human eye is the result of a long evolutionary process.

Quite apart from Darwin's explanation of the causal question, however, there was a fundamental objection to the use of even so apparently wonderful an example as the eye as the proof of God's will and purpose.[57] If one were designing such an eye from scratch, obviously the nerves leading to the brain would come straight out of the back of the retina – that is how things are arranged in the eye of a common squid or octopus. But detailed anatomical investigation reveals that in every vertebrate eye, including, crucially, that of humans, the retina is back to front. The wiring (the nerves) comes off the front of the retinal receptors and is collected together within the eye and passed back through a hole in the retina to form the optic nerve. This hole, the place where the nerves exit the back of the eye, forms a blind spot with no receptors in it at all. The celestial craftsman who contrived this eye has made a jury-rigged sort of affair. Writers like John Ray made much of the fact that this blind spot is not placed dead centre in the retina, but off to one side, where its presence is less intrusive. But that begs the question of its existence, and the reversed wiring pattern, in the first place.

The eye is not just imperfect, it is illogical and a telling example of what Hume called the 'inaccurate workmanship' of nature. No one designing an eye would have put the nerves in front of, and partially blocking, the receptors. The evolutionary explanation of the inside-out structure of the vertebrate retina is that it is a consequence of the way in which the eye develops, and that in a real sense the eye is self-organising.[58] The eye arises in the embryo first as an interaction between the outer wall of the developing brain (including the light-receptor cells that will form the retina) and a concentration of cells in the outer 'skin' of the embryo. The latter will form the lens. The nerve cells in the early outer wall of the brain have nerve fibres oriented outwards. When this wall becomes folded in to form the retina, the nerve fibres are in the wrong position. Thus the structure of the eye is the logical consequence of the (lawful) way in which it forms. The same laws that direct the folding of sheets of epithelial cells to make an eye control the folding of a piece of cloth or the skin on the surface of a custard. The particulars of the vertebrate eye are therefore due entirely to Second Causes. If, on the other hand, the eye had been made by the First Cause, he would surely have got it right. If humans are God's ultimate creation, why did he design for us such an imperfect eye, especially when, if the biblical sequence of creation is correct, he had on the day before endowed some sea creatures with a far better one. The only argument to the contrary is the one pointed out by Richard Bentley in 1692: 'If the

Eye were so acute, as to rival the finest microscope . . . it would be a curse and not a blessing to us.'[59] But that will not do.

One does not have to look far for other cases of the imperfections of creation. Many humans suffer from back pain at some time in their lives. Part of the cause may be the inconvenient fact that the human sacroiliac region, where the lower backbone meets the pelvis at the sacrum and where the whole weight of the body is borne upon the legs, is mechanically imperfect for sustaining that weight. The result is that the cushioning discs between the lumbar vertebrae may be pushed out of position or compressed or may actually crumble, and the nerves, which pass out between the vertebrae, then become pinched – painfully so, the pain being felt along the line of the nerves, the most common being sciatica in the legs. One explanation of this is that God created us this way; presumably our suffering builds character. Another explanation is that, in our evolution from four-footed forms where the weight was distributed between the fore and hind limbs, we have not finished the business of making a good new mechanical arrangement; we place too much pressure on this region at the wrong angle.[60]

A final example of the imperfection of humans exists in the fact that a large portion of the (Western) human population has trouble with its wisdom teeth. The teeth of the human jaw erupt in sequence from front to back in two waves. The incisors, canine and premolars first appear as milk teeth which are progressively lost and replaced by the permanent teeth. This allows us to have a good set of smaller teeth while we are still growing (teeth do not grow). As the wave of replacement proceeds, the molars begin their eruption, starting with the first molar. But the molars have no milk tooth in advance to secure a place for them. The last molar of all is the so-called wisdom tooth and does not erupt until the teenage years or later, by which time the jaw itself has stopped growing. In many people, there simply is not enough room along the jaw line for the wisdom teeth to make their way in. The result is 'impaction'. The scientific explanation for this is that the modern human head, with its very large cranial vault and forwardly directed eyes, is still changing and has created a much-shortened jaw compared with our ape relatives. As a result, we are in the process – captured at this moment in time – of evolving a shorter tooth row and losing our wisdom teeth.

Parenthetically, there is an evolutionary dilemma here. A Darwinian explanation of loss of wisdom teeth would include the added factor that, while the jaw was undergoing an evolutionary shortening over many generations, individuals with badly impacted wisdom teeth, leading to terrible pain and infection, would be less 'fit' and have fewer offspring, hastening the process of change. However, the intervention of dentists means that those with impacted wisdom teeth are no longer at a Darwinian selective disadvantage and the production of unnecessary wisdom teeth will continue all the longer.

Another problem relating to the erect posture of humans, which concentrates all the architecture of the body on the pelvis, is that it places limits on the size of the birth canal. At the same time, the relatively huge head of the human foetus (due to our evolving increase in relative brain size) makes the very process of birth difficult in many cases. There is a nice balance to be struck here between the optimum brain size at birth and the optimum hip configuration.

All of nature, and certainly humans, looked at in this way, is revealed less like some system of perfectly designed form and function and more like a work in progress. Paley's proposition about a watch – that it need not be perfect to demonstrate purposeful intelligence – fails here and we seem to be revealed either as God's mistake or as nature in transition. Paley, however, was prepared to reach far beyond the sort of evidence that John

Ray had adduced for natural theology; he attempted a rather more difficult argument involving the properties of life itself.

Modern anti-evolutionists confidently assert that 'Paley's famous first paragraph concerning the watch is exactly correct' and it is of course true that a machine like a watch requires a designer.[61] But whether the syllogism is correct (a simple categorical syllogism) and an organism also requires a designer is not something to be asserted but something to be proven. Paley knew that much more would be needed in order to make his case than just the original watch analogy and massed data on the adaptations of organisms. He knew he had to ask the key question: What distinguishes a living organism from a machine like a watch? Because obviously something was wrong with, or at least missing from, his famous watch analogy. A watch is not alive; it has no capacity to sustain its own operations. While a living thing can be thought of as a complex machine, even with the most generous interpretation a complex machine cannot be thought of as a living thing. 'Living' is therefore something set apart. This meant that Paley, although not a scientist, was forced to probe into the biological properties of nature and into life itself in order to establish that it was God-given.

The essential core of creation is not the diversity of animals and plants, but life itself. Life is a double-barrelled phenomenon. In one sense it is something episodic: we are born, we live, we die. Birth and death represent the ultimate mysteries: it is not for nothing that we still refer to the 'miracle of life' when a baby comes to life in its mother's womb and in her arms. Doctors, lawyers, clerics and philosophers have argued about when life starts: but obviously the new life, each new entity, begins in a biological sense at the moment that a new nucleus is made in the egg cell by the union of the chromosomes from egg and sperm. Once those two half-cells from different parents become one cell, a new life has begun. The germ cells – sperm and egg – cannot exist on their own except momentarily; they cannot reproduce themselves and therefore represent a sort of shadowy half-living gap between the two parents and the one offspring: a discontinuity.

Life is also continuous, both in the sense that all living organisms have the same defining properties, and in the sense that all life proceeds from existing life. The life that lies for the first time in its mother's arms is part of a continuum stretching back to two parents, four grandparents, eight great-grandparents . . . all the way to the origins of life. Whether we believe that God created life or that life evolved out of non-life, whether we believe that the earth is 6,000 years old or 4½ billion, this unity and continuity of life remains. However, the idea that life is a continuum is relatively modern. Until the mid-eighteenth century (and indeed later), many people believed that new life was created out of nothing all the time. They called the process 'spontaneous generation', which had long seemed to be one of those common-sense 'facts'. If you leave a bowl of milk or broth long enough, moulds and even little creatures will appear in it. If you steep some hay in water, dozens of transparent microscopic animals will begin to swim around. There was even a name for such creatures – 'Infusoria'. Similarly, meat left out in the air will 'develop' maggots.

With spontaneous generation, the creation of life out of non-life was therefore nothing particularly special, which is interesting given that it is so obviously at odds with the biblical story of creation, in which God made life just once. We now know the old evidence for spontaneous generation is misinterpreted: the broth and the infusion of hay are merely convenient cultures for the spores and eggs of existing organisms. Maggots only appear in meat if flies are allowed to lay their eggs there. Although we associate progress in this field with Louis Pasteur

(1822–1895) – whom we have to thank for pasteurised milk, contributions to the science of wine-making (which begins with 'spontaneous generation' of yeasts in a culture of grape juice) and for the concept of germs in medicine – the first break-throughs in this area came from two Italians, Francesco Redi (1621–1697), who solved the fly-egg-meat-maggot problem, and Lazaro Spallanzani (1729–1799), who discovered that any culture heated to 100 degrees centigrade for long enough would be sterile. By 1800, the notion of spontaneous generation of lower forms of life had been quietly discarded, if not decently buried. This suited Christianity perfectly, because it reinstalled creation as a single initial divine act rather than a repetitive process. For the natural theologian, non-living matter must differ in some way from living matter. There were only two possibilities: that life is a property of matter when the right atoms are in appropriate motion, configuration and environ-ment, or that there is some extra-material 'vital force' that animates non-living matter. A nice irony had been produced: science, having debunked spontaneous generation, soon found itself needing to postulate it again in the form of a single mole-cular event of self-assembly at the start of life.

Those who live by the sword, die by the sword. The sword of all natural theology was rhetoric and Paley, committed to his central analogy, was forced to plunge on with it, into an argument about the processes by which life itself has constantly to be recreated in the cycles of birth and death. Given the philo-sophical minefields involved, he might have been expected to do exactly the opposite and, like John Ray, quickly move to present a myriad more documentary evidences of the complexity and designed-ness of nature. Not so William Paley:

> Suppose, in the next place, that the person, who found the watch, should, after some time, discover, that, in addition to all the

properties which he had hitherto observed in it, it possessed the unexpected property of producing in the course of its movement, another watch like itself; (the thing is conceivable;) that it contained within it a mechanism, a system of parts, a mould for instance, or a complex adjustment of laths, files, and other tools, evidently and separately calculated for this purpose; let us inquire, what effect ought such a discovery to have upon his former conclusion?

In order to extend his analogy, Paley had now redefined his watch as an impossible machine that contained a self-assembling device. Evidently this hypothetical watch, with its internal workshop, only needed to take from its environment some raw materials in the form of brass, silver, steel spring and so on in order to fabricate another watch. On the face of it, this 'system of laths and files' is ludicrous, and the sceptic's immediate response will have been to point out that it would have to be far more complex than the watch it was hypothetically capable of making. Even to this day, humans have not succeeded in creating self-replicating, self-assembling machines, although there is evidence that this can be done in the digital, electronic realm of computers. But it is actually quite prescient; the system of lathes and files is a great metaphor for what living organisms actually do. Paley was ignorant of how they did it; he knew it happened, but preferred to ascribe the whole mechanism to God's direction. But in the cells of living creatures there is indeed the wherewithal to make another organism. The machine needs an external blueprint and a fabricator; the living organism has its *own* blueprint in the form of its DNA: it is its own blueprint and is the impossible machine that self-assembles. And there's the rub: as it designs and redesigns itself through trial and error, it needs no designer.

Even on its own terms, this metaphor of lathes and files led

Difficulties with the Theory, and the Argument Extended

Paley into all sorts of trouble concerning the concept of 'cause' – an essential battle for him to join. The person who found the metaphorical watch on the heath, when he discovered that it could spawn further watches, would reflect that 'it was in a very different sense from that, in which a carpenter, for instance, is the maker of a chair; the author of its contrivance, the cause of the relation of its parts to their use'. With respect to these, the first watch was no cause at all to the second. 'The watch is no more the cause of another watch' (even via its hypothetical system of lathes and files) than the 'mill is the cause of the flour'.

This is cause in one of the four senses that Aristotle used the term: a chair has a material cause in the wood from which it is made; a chair made from paper would be different. There is an efficient cause in the form of the carpenter who made the chair. There is a formal cause in the design from which the carpenter proceeded. And there is a final cause in the purpose for which the chair is intended. In using Aristotle's categories, however, Paley was pulling the wool over his readers' eyes: the chair does not make another chair. When a carpenter makes a chair, he is only the efficient cause, not the formal or material cause. Even in Paley's time, in the absence of spontaneous generation, living organisms had to be seen as their own material, efficient and formal cause (even though the mechanisms and natures of those causes were unknown). Only the theologians' First and Final Causes were a matter of debate.

More immediately, there was a second philosophical reference problem. When Paley said that 'the first watch was no cause at all to the second', he was tangling with yet another of the irresistible brain-teasers that Hume posed in *Dialogues Concerning Natural Religion*. Again it starts with Aristotle, who had proposed an eternal world in which there was no beginning and therefore, implicitly, no First Cause. Obviously, theologians

101

and natural theologians of the age insisted the contrary. Samuel Clarke, in the first Boyle Lecture, pointed to the 'absolute impossibility of an Eternal Succession of Dependant Beings, existing without any Original Independent Cause at all ... Something must needs have been from Eternity.' David Hume, expanding the debate, asked whether this was true. Was it possible that in a long chain of cause and effect the cause of any one link could also be the cause of the whole chain? If each living thing is the cause of the next, what need is there of any external agency? One obvious answer is that such a chain of life still needs an initial cause (unless of course life is a circle, not a chain, in which case one needs an explanation for the whole circle). Each living organism is effectively created by a preceding one (usually a pair), just as Paley's watch was projected to make a new watch. So Paley felt confident in asserting:

[Nothing] is gained by running the difficulty further back, i.e. by supposing the watch before us to have been produced by another watch, that from a former, and so on indefinitely ... a chain, composed of an infinite number of links, can no more support itself, than a chain composed of a finite number of links ... by increasing the number of links ... we make not the smallest approach, we observe not the smallest tendency, towards self-support.

Looking at this argument from a twenty-first-century point of view, one difficulty is obvious. Paley's 'watch', as everyone would agree, was not created *de novo* by a human designer. Indeed, it is hard to think of any machine of human devising that has not gone through a period of development and improvement. Paley's watch was the result of a long history. It traced back its origins to early people noting the regularity of the (apparent) movement of the sun across the sky, to the sundial, the hourglass, the water-clock, to clocks for marking the hours of prayer, to clocks with minute hands, to marine chronometers. And throughout this genealogy of experimentation, blind alleys, breakthroughs, new ways of accomplishing the same goals, other inventions allowing new functions – in the great sweep from weight to pendulums, to springs (and eventually to quartz crystals and even atomic devices) – there is a nice analogy for the evolution of life on earth.

Paley did not know what we know. Even so, he was well aware of the fact that everything would be different if the chain of history were one of growing complexity, starting in some non-living system and proceeding to the diversity of life in all its infinite variety by utilising what he would have called Second Causes. A theory of change, for example, would envisage the chain of life as one beginning from the 'chance collisions of atoms' and passing through lower animals, eventually to man himself. Such theories were already in existence in 1800. That was Paley's terrible dilemma. People such as Erasmus Darwin and a host of materialist philosophers of the eighteenth century (mostly from the Continent) had long since proposed such hypotheses. They simply could not confirm them with evidence. Therefore here, as elsewhere, Paley was really fighting a rearguard action, and depending on there being no middle ground between a theory of creation and a theory of change. As the Reverend Ralph Cudworth had written: 'there is no Middle betwixt these Two; but all things must either spring from a God, or matter . . . Nothing from Nothing Causally, or Nothing Caused by nothing, neither Efficiently, nor Materially.' If everything, each species, is fixed for eternity, there is no middle ground. If they can change, the middle ground becomes a reality.

Paley launched the lathes-and-files metaphor knowing that it might have just the wrong effect and might instead intrigue his readers to the contrary view. Having envisaged a watch that could make a watch, he opened up the possibility that there is no longer a need for a watchmaker. There remained only the argument of last resort: that the discovery of a system of lathes and files – the equivalent of discovering all the secrets of the processes by which living things reproduce – would only 'increase his admiration of the contrivance, and his conviction of the consummate skill of the contriver'. Sixty years later, the Reverend Charles Kingsley would echo this exactly: '[it is a] noble conception of deity to believe that he created primal forms capable of self development.'

Typically, Paley tried to herd his readers in a single direction:

> The watch in motion establishes to the observer two conclusions . . . that thought, contrivance, and design have been employed in forming . . . its parts . . . [and] that force or power, distinct from mechanism, is at this present time acting on it. It is the same in nature. In the works of nature we trace mechanisms; and this alone proves contrivance but living, active, moving, productive nature, proves also the exertion of a power at the centre.

As to this power, he likens it (in a watch) to 'a secret spring or a gravitating plummet; in a word there is force and energy, as well as mechanism'. That secret spring is God's own and unique vital power.

The very language used here reveals Paley's familiarity with the 'atheist's' completely opposite version of a machine analogy. In 1768, the French materialist philosopher Julien Offray de La Mettrie had written a book that encapsulated the whole problem in its title – *L'Homme machine* – and in one simple sentence. A living organism is totally different from any machine because the body is, in effect, 'a machine that winds itself up, a lively piece of perpetual motion'.[62] Life is simply a property of the materials from which it is composed; no external vital force is needed to wind it up or to start it ticking.

Again Paley has to resort to assertion:

Mechanism is not power. Mechanisms, without power, can do nothing. Let the watch be contrived and constructed ever so ingeniously . . . it cannot go without a weight or spring, i.e. without a force independent of, and ulterior to, its mechanisms . . . The intervention and disposition of what are called 'second causes' fall under the same observation . . . Neither mechanism, therefore, in the works of nature, nor the intervention of what are called second causes, (for I think they are the same thing,) excuse the necessity of an agent distinct from both.

There was an additional hazard here. The concept of the living organism as a wholly material machine begged a question that brought the whole discussion dangerously close to home. If all life is material, what is the mind and what is the soul? The body can only be a machine, Descartes had argued, if the soul is situated, and owes its causal nature, elsewhere. Throughout the eighteenth and nineteenth centuries, bitter arguments raged about such a dualist compromise. Materialists such as La Mettrie dared to argue that the mind and soul were simply the product of the material properties of the body. Their opponents in turn denounced the notion that 'reflection, judgement, memory, arise out of changes . . . we call chemical' or that 'an immaterial and spiritual being could . . . have been discovered amid the blood and filth of the dissecting-room'.[63] In 1827, Charles Darwin's second year at medical school in Edinburgh, a fellow student, William Browne, gave a presentation to the Plinian Society (a student scientific club) on the material basis of the mind. His account was expunged from the minutes of the meeting.[64] Like most natural theologians, Paley did not venture into this particular debate. Nowhere in *Natural Theology* did he examine the nature or source of the mind and soul. For him it was as obvious that they are God-given as that a watch serves to tell the time.

Ironically, a machine like a watch does not suffer from a particular property of living systems. It is potentially – through human intervention – immortal. Each gear and spring can be replaced until it becomes like George Washington's axe, the head of which has been replaced twice and the handle three times – but it remains George Washington's axe. Living systems are not so infinitely reparable.

Whatever the lathes and files metaphor was meant to accomplish, it definitely was not meant to open the door to the materialist idea that living organisms are self-assembling. For Paley, whether in the maintenance of life in a single organism, or the generation of one organism from another, or the creation of the very first life, the power of God was needed: 'Whatever includes marks of contrivance . . . necessarily carries us to something beyond itself . . . to a designer prior to, and out of, itself. No animal, for instance, can have contrived its own limbs and sense, can have been the author to itself of the design with which they were constructed. That supposition involves all the absurdity of self-creation, i.e. of acting without existing.'

In 1744, however, a Swiss zoologist, Abraham Trembley, had shown that this was not so absurd after all. In freshwater streams the world over, one can find, attached to stones, a tiny transparent creature looking like a minute version of the trifids that terrorised the world in John Wyndham's brilliant science-fiction story. This little animal is called – aptly, as we will shortly see – *Hydra*, after the monster of Greek myth who grew two heads where one was cut off, and is a relative of the larger, more complex and highly coloured sea anemone. If any one fact slayed the hypotheses of the natural theologians, it was the observation of how a simple hydra conducts its vital zoological business. What seems one of the lowest forms of life has a great deal to tell us about life itself. Trembley's experiments had been known for fifty years before Paley wrote, but natural theologians had closed their eyes to a whole range of awkward facts such as this.

Trembley discovered that, if you take a little hydra (or 'polyp') and cut it into pieces, each part will grow into a new hydra. Now it was well known that if you cut an earthworm in half, the head end would grow a new tail, but the tail could not grow a new head. The former is wound repair, the latter strays into God's territory. A hydra has no such scruples: cut it into small pieces and each will grow into a whole animal.[65] On investigation, a host of other lowly creatures had similar properties; a flatworm (not a true worm) will grow two perfect heads if you cut the first one down the middle. The animal can, after all, directly be 'the author to itself of the design with which they were constructed'; it *is* a mechanism that metaphorically winds its own spring as it makes itself from simple foods, water and air. The hugely complex 'machine' that is the human body makes itself out of a single fertilised cell. As these are animals, not mere plants, this even opens up the question of whether each has the capacity to make its own soul.

In 1800, all Paley could do was insist that none of this was relevant: '[This does not] in any wise affect the inference, that an artificer had been originally employed and concerned in the production.' He has no recourse but to insist on one argument, that of Final Cause, or purpose:

The question is not simply, How came the first watch into existence ... This, perhaps, would have been nearly the state of the question, if nothing had been before us but an unorganised, unmechanised substance, without mark or indication of contrivance. It might be difficult to shew that such substance could not have existed from eternity, either in succession (if it were possible, which I think it is not, for unorganised bodies to spring from one another) or by individual perpetuity. To suppose it to be so, is to suppose that it made no difference whether we had found a watch or a stone. As it is, the metaphysics of that question have no place; for, in the watch which we are examining, are seen contrivance, design; an end; a purpose; means for the end, adaptation to the purpose . . . the thing required is the intending mind, the adapting hand, the intelligence by which the hand was directed.

This, however, only begs the question, which really is: 'How came the first watch into existence?'

Opening up the lathes-and-files issue had highlighted the problems of how any living thing comes into existence, how the first one did, and all the causes of what lay between. This was to become the pivotal set of issues threatening natural theology. New discoveries in biological processes and the search for the causal mechanisms of life itself – Second Causes – would threaten to undermine the First Cause; and that was atheism.

As early as 1693, for all that he carefully accumulated examples of the marvellous and intricate structures and mechanisms of nature, Ray had clearly seen the dangers – the signs of atheism – ahead. And the dangers did not lie only with the limitations of the argument from design or the cleverness of philosophers, or even in the fact that contemporary understanding of the material processes of life was so incomplete. The dangers lay in philosophies of change having to do with the nature of the physical earth, rather than the living nature that occupies it, and arose primarily from the work of geologists. By Paley's time, the problem was acute. Many serious seventeenth-and eighteenth-century scholars had been concerned with finding out the hows and wheres, and even the whys and whens of the earth's structure, and the study of geology had begun to reveal a whole series of new kinds of truths. The earth's surface was found to be a complex structure and its constituent rocks, whether deep in the earth or at its surface, belonged to a finite number of recognisable types. A rational person was entitled to ask the questions that had not previously been allowed: Why is the earth not uniform? Why are bits of it folded? Why are some bits white chalk and others red, grey or green sandstones? Why are some hard granites and some soft talcs? Soon, it was possible to ascribe (sometimes incorrectly) a secondary cause to each of these rock types. And all of the causes involved change. Some rocks are crystalline and have been subject to intense heat and pressure. Some are clearly formed by deposits of sediment in the form of sand, silt and clay. These sedimentary rocks seem to have been formed in water, their constituent particles moved by water and indeed eroded from older rocks by water. Some rocks have been formed by volcanic activity. Not only could all these patterns and processes of change be inferred from the structure of the rocks, they could be tested by observation and even primitive experiment.

Obvious to anyone who would see, and even more apparent to anyone who would look closely, was the fact that the structure of the earth is also a complex system, consisting of many different kinds of materials and structures which are strictly and predictably patterned and therefore lawful. One does not find fossils in granite or emeralds in sandstone. The earth is built up in most places into layers. Sometimes these layers may be seen lying horizontally, like the pages of a book on its side, sometimes they are thrown in complex folds like bed clothes after a restless night. Sometimes the layers stand vertical. Some constituent layers have obviously been formed in molten state and then quickly cooled, some layers show ripple marks identical to those seen on any beach. And many contain fossils. Then, when geologists started to piece together the complex nature of the earth's crust, they realised that underneath the surface rocks was a vast, super-hot, perhaps even molten, interior – of the sort of material that occasionally spews out of volcanoes and the heat of which was evident when one descended a mine. Whereas the Church had for more than 2,000 years taught that God created the earth as it is now, in a single event, the new picture was one not only of a complex system, but an ancient changing one.

CHAPTER SIX

Fossils and Time: Dr Plot's Dilemma

'When the coherence of the parts of a stone [are] so inexplicable . . . with what assurance can we decide concerning the origin of worlds, or trace their history from eternity to eternity.'

David Hume, *Dialogues Concerning Natural Religion*, 1777

'The vulgar and inconsiderate Reader will be ready to demand, What needs all this ado? To what purpose so many Words about so trivial a subject? What Reference hath the Consideration of Shells and Bones of Fishes petrified to Divinity?'

John Ray, *Three Physico-Chemical Discourses*, 1693

Given the fact that William Paley must, from his thorough reading of contemporary science, have been aware of all the developments in geology and palaeontology and their challenges to biblical authority, the wording of his version of the watch analogy in the first paragraph of his book presents a puzzle. For all the brilliance of the argument, there is something curious about what Paley had added to the discovery of the watch: 'In crossing a heath, suppose I pitched my foot against a stone, and were asked how the stone came to be there, I might possibly answer that, for any thing I knew to the contrary, it had laid there for ever.' The more one reads it, the more a nagging hint of contradiction appears and there seems a strange defiance in the rest of the sentence: 'nor would it perhaps be very easy to show the absurdity of this answer.'

A stone: on the face of it he is merely giving us something commonplace and uninteresting to contrast with the gleaming mechanical wonder of the watch. But stones can be simple and, as Hume had observed, they can be complex: marbles like those carved by Michelangelo, gleaming chalk, deep blue slates, useful flints, precious stones like rubies, semi-precious opals and garnets, ores of gold, iron or coal – 'stones' are far from dull or insignificant. And Paley never used words or analogies casually. Was the stone, seemingly so readily dismissed, nothing to Paley or was there, in his studied indifference, a metaphor after all that he meant us to read? It is possible that, when he said 'for ever', he might simply have meant 'since the third day of creation'. But in saying that it would be difficult to prove that it had *not* been there for ever, he seems discreetly to be alluding to the contemporary geological debate over the age of the earth. Certainly there was a direct connection to the words of David Hume, quoted above, and we may even ask whether there is a coded reference to something else. After all, it was a stone that Dr Johnson kicked away to demonstrate the absurdity of Bishop Berkeley's notion of the non-existence of matter. It would perhaps be too far a reach to see a biblical pointer to the philosophical dangers inherent in the stone, as in the words of the Psalmist: '[His angels] shall bear thee up in their hands, lest thou dash thy foot against a stone', the very words with which the Devil tempted Jesus in the wilderness (Psalms 91:12; Matthew 4:6). One way or another, an awful lot is potentially bound up in the apparently simple choice of 'a stone' in the watch analogy because, if not from the hand of God, the stone must have come from some other cause.

As seventeenth- and eighteenth-century geologists delved into the nature of rocks and the complexity of the strata in which they are arranged, they also speculated about potential causes of change in the earth's crust. Among the earliest of these contributors to geological science was Robert Hooke, who, with a small number of like-thinking scholars, produced a growing mass of evidence that the earth had changed over aeons of time and in ways not discussed in biblical narratives. In 1668 Hooke read before the Royal Society his lecture entitled 'Discourse of Earthquakes',[66] where he produced a very modern geology in which the complex structure of the earth's crust is due to the violent actions of volcanoes and earthquakes. In his view also, fossils are simply the remains of real creatures, long dead, and the dry land where they are found was 'once cover'd with the Sea [and] . . . had Fishes swimming over it'. None of this could be evidence of 'the Flood of Noah, since the duration of that, which was but two hundred Natural Days, or half an Year, could not afford time enough for the production and perfection of so many and so great full grown Shells.' At subsequent meetings of the Royal Society he added various notes and observations on the subject – concerning fossils as much as earthquakes – although they remained unpublished until the posthumous collections of his works edited for the Royal Society by Richard Waller in 1705. One year after Hooke delivered his seminal lecture on earthquakes, a Danish scientist named Niels Stenson or Steenson (and like Linnaeus, universally known from the Latin version of his name: Steno) set down some fundamental rules for understanding the changes demonstrated in the patterned ordering of the rocks in an extraordinary little essay. One of these rules, Steno's principle of 'superposition', was breathtakingly simple: beds closer to the surface must have been laid down more recently than the ones underneath, and the lower beds did not interfere with the formation of the ones on top. It is so obvious that one asks why no one had articulated this before. Armed with such simple keys, geologists could unlock more secrets of the earth.

But one thing was missing: time was the geologists' dilemma; everything that the geologists had discovered about geological processes required time, and they needed more than the Bible allowed. All the new evidence of the changing complexity of the earth posed questions for the Bible that simply could not be ignored; it presented the most powerful and direct challenge imaginable to the account of creation in the book of Genesis. The structure of the earth did not look as though it was formed in six or even a thousand days; the processes by which the earth appeared to change pointed to a history of hundreds of thousands, perhaps millions of years for which there was no authority in the Bible. On the other hand, if time was granted – millions of years of it, during which the same processes we see acting today had carried out their inexorable microscopic actions – other difficulties fell away.

As geologists made a case for an ancient earth they created the first great empirical threat to religion from science, directly undermining the authority of the Bible and particularly its first book, in which the creating hand of God was revealed. Before the rise of geology, the threats were theoretical, philosophical, logical; after the emergence of geological science they were concrete. This new breed of scientists would run square into the problem of biblical authority unless theology could find ways to embrace and control the sciences of the earth.

The issue took a long time to resolve. Hooke was able to question the age of the earth and to proclaim the organic nature of fossils with relative impunity. The strength of clerical reaction only grew slowly, with the reality of the threat. There was an additional factor: theologians thought they had a cast-iron case – from the Bible itself – that the argument for an ancient earth was false. In retrospect, an insistence on the literal truth of the first thirty-four verses of Genesis (for the exact wording, see

Appendix A) seems unnecessary. The proposition that God had created the world and all living creatures did not depend on his having done it at breakneck speed. But adherence to Genesis had become (and remains for many) a sort of loyalty oath. And it had gathered around itself a plethora of auxiliary 'truths' which also had to be accepted – and were. Prime among these were various calculations of the age of the earth, based on analysis of internal biblical chronologies. In the critically important matter of creation, the Church felt itself reasonably secure; it had a smoking gun in the form of a historical record giving the age of the earth. The actual date of creation had been calculated.

The notion that the world was created on Sunday, 23 October 4004 BC depended principally on the kind of analysis made famous (or notorious) by Bishop James Ussher of Armagh, Primate of All Ireland, in his 1650 *Annals of the Old Testament deduced from the First Origin of the World*: 'The Julian years, with their three cycles by a certain mathematical prolepsis [which] have run down to the very beginning of the world ... an artificial epoch, framed out of three cycles multiplied in themselves: for the Solar Cycle being multiplied by the Lunar, or the number 28 by 19, produces the great Paschal Cycle of 532 years ...' and so on. This was combined with a reckoning from the genealogies listed in the Pentateuch, and the resulting chronology was appended to the King James Bible in 1701, thus ensuring a huge audience. Ussher is usually also credited with the additional proposition that creation had occurred at 9.00 a.m. in the morning of 23 October, when in fact Sir John Lightfoot Vice-Chancellor of the University of Cambridge, was the one who decided this. (Ussher had actually opted for midday.) These propositions seem to have strained the credulity of few contemporary scholars. In fact, for years they seemed to constitute the very best in precise and scientific (to say nothing of pious) scholarship.

The premise that the account of our origins in Genesis is literally true and that the events recorded in Genesis, Exodus, Leviticus, Numbers and Deuteronomy document the totality of man's early history, is not sufficient to establish a chronology. It is also necessary to assert that a biblical day was the same as a common day, that each biblical month was thirty days, and to accept that Methuselah really lived 969 years and Noah 950. If these assumptions are not made, the earth would appear either younger (if Methuselah only lived for 70 modern years, for example) or older (if the six 'days' of biblical creation were six years, or six thousand years, and so on) – this is an on-going debate. There is also the question of which translation of the biblical canon one uses: the Septuagint would have given Ussher a different date from the Hebrew Bible on which the Vulgate was based. Prior to Ussher, the date of creation had been thoroughly argued out (without conclusion) by Celsus in the first century AD, Origen in the third, Basil in the fourth, Augustine in the fifth, and so on, even before the Vulgate Bible had been set out. All of these analyses depended on a strict, internally consistent logic, even though, to our eyes, they fall under the rubric of what Jesus called 'Ye blind guides, which strain at a gnat and swallow a camel' (Matthew 23:24). And the next difficulty that had to be swallowed was the nature – even the very existence – of fossils.

If a fish gliding apparently effortlessly through the water, so perfectly matched to its surroundings, can be interpreted as a creation from God's guiding hand, what should we make of a fossil fish frozen for eternity in an ocean of chalk, or a giant shark's tooth entombed in sandstone? If a pearly nautilus with its lustrous, geometrically immaculate shell is almost a work of art, what can be learnt from a fossil nautiloid perched high on a cliff never again to be restored to the sea? In all the long history of attempts to reconcile incomplete new knowledge with inadequate old ideas, few subjects have been more controversial than fossils. They have been familiar to us since the beginning of civilisation: Xenophanes wrote about them (and even about ideas of evolution) around 500 BC. The first fossils were simply found on the earth's surface or picked out of exposed rocks (the word 'fossil' itself means 'dug up'). Later, when men started to dig into the earth in earnest to make roads, canals and mines, fossils began to tumble out of the rocks – a cornucopia of 'figured stones' or 'formed stones' apparently resembling bones, plants and, above all, shells.

Nowhere on the surface of the earth is one far from being able to pick up a fossil. On many a heath the stone against which any latter-day Paley pitches his or her foot will contain a fossil. Even in Antarctica, wherever rocks emerge from the ice, there is a chance of finding fossils – dinosaurs, great ammonites, ferns and palms – all evidence of tropical climates and long-extinct creatures where now there is but bitter cold. But while fossils are familiar to us and we generally understand their origins, in earlier times they were not only an enigma but a threat. They have many of the appearances of living creatures and yet they are made of 'stone'. This forced the formulation of critical questions: If fossil shells were once the shells of real marine creatures, now encased in and turned into stone, how long ago did that happen and how did they come to be entombed deep under the earth or high up in mountains? If they are merely 'figured stones' that have a coincidental resemblance to real organisms, what caused them? If God had made the earth as it is now, why did he fill it up with these odd 'figured stones' and did he do so on the third day when he made the earth, or on the fifth day when he made all the sea creatures? Or do fossils represent the result of some odd Second Cause? (But then again, what?) And who dare ask whether, buried in the rocks, there might be human fossils?

Equally puzzling was the fact that fossils seemed to be in the wrong places. These strange 'formed stones' mostly had the appearance of sea creatures, but they were found on land. In fact, the best place to find them seemed to be mountainsides. If the only logical possibility was that they had been caused by the Flood, then they were simply the remains of creatures caught up in the debris of that great inundation and cast about into odd places like mountain tops. But that explanation was too simple. As many commentators pointed out, if fossils represent the remains of real sea creatures killed by the Flood, their remains would have been washed to the bottom of the ocean never to be seen again.

Perhaps the trickiest problem posed by 'formed stones' was that most of the creatures they appeared to represent were missing from the earth of today and must, then, represent species that had become extinct. But there is no biblical authority for that; the account of creation in Genesis ends with God pronouncing that 'it was good'. It does not leave open the possibility of correcting mistakes. The story of Noah (both versions) shows that all the animals were saved at the time of the Flood – it was the wicked humans who were culled.

In the seventeenth and eighteenth centuries, as collecting fossils for 'Cabinets of Curiosities' became popular among academics and learned amateurs, a number of challenging facts began to emerge. It became clear that different fossils are found in different layers, distinctly and predictably patterned through the earth's crust. Many kinds of rock (granites, for example) held no fossils at all. Fossils only seemed to be found in what are now called sedimentary rocks, laid down in water as discrete strata of mudstones, limestones and sandstones. But, if fossils represented creatures drowned in a flood and buried in diluvial mud, there must have been successive floods in order to account for the patterns in the rocks. That in turn would imply multiple creations and a constant process of restructuring the earth. Finally, fossils seemed to confirm the purely geological evidence of great antiquity for the earth, making such rebuildings more feasible. One could not simultaneously hold the view that the world and all living creatures were created 6,000 years ago and hold that fossils and the general geological story – which spoke of (at least) hundreds of thousands of years of change – were a true record of the history of the earth. A great deal depended on the correct interpretation of these 'formed stones'.

Those early days of the modern age of fossil-hunting were full of interesting characters such as Jacob Scheuchzer (1672–1733), who was perhaps the first to have thought he had actually found a human fossil, a victim of Noah's Flood. But his '*Homo diluvii testis*' ('man witness to the flood') turned out to be a fossil giant salamander. (If nothing else, this in turn was the inspiration for Karl Capeck's classic science-fiction novel *The War with Newts,* in which a living salamander mutates into a humanoid.) But among all the characters from the seventeenth century who anguished over the problem of fossils, one of the most informative is Dr Robert Plot, who was born in 1641 (a year before Newton) and died in 1696, thus living through one of the great periods of English science. Writing in 1677, he gives us a view of the debates that led to modern views of the age and history of the earth, and a very different explanation of the origins of life on earth.

Plot was born into an established, land-owning family in Kent, where his father had been Captain of the militia of the 'hundred of Milton'. He entered 'Magdalen Hall', Oxford, earning his MA in 1664, and at the age of thirty became Doctor of Laws just at the moment that Oxford science was really flowering under Wren, Boyle, Hooke, Lister and others. Plot made a name for himself as a student of antiquities; like John Ray he was a collector of objects and of facts, but lacking Ray's

Robert Plot

discernment, who found a new natural order of things, he saw himself simply as a chronicler. While Newton, Boyle and others belonged to the new science that looked for the material causes of phenomena, unhampered by the weight of ancient authority, Plot belonged among the last of the old school of natural philosophers. And it is precisely for this reason that he earns a place in this story; in his writings he captures the dilemmas of the old school when faced with the new. Plot was a man torn between conflicting styles of explanation and of science, between looking backwards to classical authorities and daring to follow more modern scientific ideas to their logical conclusions. We see it in all his work and particularly as he wrestled with the subject of fossils, something that both fascinated and troubled him.

At the age of thirty-six, Plot published his magum opus, *The Natural History of Oxford-shire* (1677), which he followed in 1686 with *The Natural History of Stafford-shire*, intending eventually to write an encyclopaedic account of the natural history of all England.[67] He died before he could write any more, but these two form a more than adequate memorial. Both were probably modelled on Joshua Childrey's *Britannica Baconica or the Natural Rarities of England, Scotland and Wales* (1661), a sober gathering-together of the facts. In similar Baconian fashion Plot assembled the raw data and used it to draw conclusions, rather than the other way around. (Among others, the Reverend Gilbert White, another Oxford man, seems to have used Plot's *Oxford-shire* as a model for *The Natural History and Antiquities of Selborne* a hundred years later.[68])

Plot was not a cleric and, although no doubt a devout man in his way, he had no evident interest in physico-theological matters. He noticeably referred to fossils as the works, the productions, of 'Nature' rather than the handiwork of God. By this, nature was personalised as a force and a power, and by ascribing causes and effects to 'Nature', Plot followed the practice of his contemporaries and referred everything to Second Causes. Nature is thus the sum of all the properties and lawfulness of living and non-living matter. Plot's contemporary Robert Boyle put the matter this way: 'Sometimes when it is said that Nature does this or that, it is less proper to say that it is done *by* Nature, than that it is done *according to Nature*. So that Nature is not here to be looked upon as a distinct or separate Agent, but as a Rule or rather a System of Rules according to which those Agents and the Bodies they work on are, by the Great Author of Things, determined to act and suffer.'

In 1683, a vast collection of natural-history specimens originally brought together by John Tradescant the Elder (? –1638) and his son John Tradescant the Younger (1608–1662) and exhibited in London as 'The Ark', came to Oxford. They had been willed by the younger Tradescant to his friend and neighbour Elias Ashmole, collector and antiquary. In turn Ashmole bequeathed them all to the university along with his own collection of coins and antiquities. The university formed a great museum to house them – 'a Repository of Natural Curiosities and Antiquities in said University' – open both to students and scholars and to the public, the first of its kind in the world. Plot was made the first Keeper of the Ashmolean Museum, and at the same time 'Professor of Chymistry'. When he died in 1696 he in turn left to the museum the collection of 'Natural Bodies' from his lifetime of chronicling the objects of nature and material culture.

When Plot described the 'natural history' of Oxfordshire, he tackled everything from archaeology, echoes, lightning, historical artefacts, inventions, geology, epidemics, minerals and natural resources (animals and plants), to the local industries of glass-making and glove-making, to folk-lore and folk-medicines, improvements to navigation and the marvels of modern architecture. (One of his more fascinating accounts is of two women, Anne Green in 1650 and 'Elizabeth' in 1658, who were hanged for murder but survived the experience.[69]) And Plot described the fossils of Oxfordshire, his dilemma being that Oxfordshire was full of fossils, but he didn't know what they were. In reading Dr Plot, one can sense his discomfort as he, a good scholar, explored the nature of 'figured stones' and tried to find some explanation of their possible causes. To his credit, he tried to use a quasi-scientific method by first organising the phenomena into logical categories. But in such a venture one must either be clever or lucky in getting the criteria right in the first place. Plot chose to order his fossils into groups according to what he guessed was their cause – a fatal flaw if he intended, as a result of the exercise, to *discover* any cause.

Plot attempted to define three categories of what we call fossils: 'petrifactions and incrustations', 'formed stones' and 'nature imitating art'. The first of these is straightforward: throughout Britain one finds springs, arising out of limestone rocks, that contain a saturated solution of calcium carbonate. Wherever the water drips onto something, the calcium carbonate is deposited. In a cave, this produces stalactites hanging from the roof and stalagmites accumulating on the floor. If any object is placed in the water of such a spring for long enough, it will become encrusted with limestone. Plot concluded from such objects that 'one may plainly see how the stony *Atoms* have intruded themselves . . . into all Parts alike' or have formed 'the meerest Incrustation'. William Buckland, Reader in Mineralogy at Oxford in the 1820s and 1830s later made a large collection of experiments at these petrifying springs. Some of them still exist in the Oxford University Museum collections: birds' nests, a stuffed blackbird, a ram's head, a crayfish, and even a human skull; all apparently turned to stone. But not really; the original object still exists inside, having merely been covered with mineral, not replaced by it. Entirely different processes are involved when a true fossil is formed by the systemic 'intrusion' of those 'Stony Atoms'.

Next Plot brought together 'formed stones', that is, those not 'made . . . by the Tool of the *Artist*' (hand-axes and arrow heads, for example). At least some fossils, Plot realised, had been '*naturally Formed*, and seem rather to be made for [man's] *Admiration* than Use' and in fact it is in this category that we find the vast proportion of what we today call fossils – ammonites, fossil shells, bones and plants. But Plot also managed to include a great deal that was not fossil at all. He tried to guess their causal origins from their shapes and appearances, dividing his formed stones into three categories: '[those that] relate to the *Heavenly Bodies* or *Air*; those that belong in the *Watery*

Kingdom . . . such as resemble *Plants* and *Animals*; [and] lastly such *Stones*, wherein, contrary to all Rule, *Dame Nature* seems to imitate *Art*.' Here we see perfectly the chicken-and-egg dilemma of all early attempts at classification. It is hard to organise things without knowing their underlying causes: the causes cannot be discovered if the data has not been organised into the right categories. In retrospect, he was doomed from that moment.

In his first group, Plot lumped a variety of stones that relate to 'the *Heavenly Bodies*' and are assumedly formed by, or in, them. Obviously this is where he put fossil starfish and the related five-rayed sea urchins. He also included stones apparently formed in the shape of the moon, various minerals (on the basis of their regular crystal structure), and fragments of coral. A sub-category included stones that had been 'generated in the *Clouds*, and discharged thence in the times of *Thunder* and *violent Showers*', among which are long, pointed objects such 'thunder bolts'. (These are now easily recognisable as the fossil shells related to nautiloids known as belemnites.) In Plot's next group 'come we . . . to such as represent [water's] Inhabitants'. Here Plot describes occurrences of fossil '*sea-Fish*' – both the shelly and finny kinds. Everything Plot describes here is something we would now call a fossil proper. He thought that all of them had been formed by some physical properties of water itself that had caused rocks to take the form of creatures that lived in the water, as indeed the same power caused living fishes to be fishes, rather than, say, elephants. In his final category (where nature is thought to imitate art), Plot described a series of plant-like remains together with some flights of fancy such as flints in the shapes of a human heart, an eye, breasts, testicles, and so on. But also included are the remains of many kinds of vertebrate teeth, vertebrae, and limb bones, to the last of which we must refer in a moment.

Fifty years after Plot had catalogued these treasures of one

English county, Johannes Beringer (Dean of the Faculty of Medicine and personal physician to the Prince Bishop of Wurzburg) was famously doing the same in Germany.[70] An avid collector of fossils, he was less fortunate than Plot in that some jealous colleagues salted his quarry sites with stones carved in 'clear images of the sun and the moon, and of comets radiant with flaming tail'. As the deception mounted, his collectors brought him 'finally ... splendid tablets ... marked with the ineffable name of the Divine Jehovah in characters of Latin, Greek, and Hebrew'. Beringer believed it all, developing theories to explain that these formed stones were great works of nature. His life's work, *Lithographiae Wirceburgensis* (1726), is now quite a collectors' item.

Of all Plot's formed stones, it was the shells and obvious vertebrate remains that taxed him the most, because they were so obviously 'real'. In his honesty, he made things more difficult for himself by rejecting the popular view that formed stones had been brought there 'by the *Deluge* in the days of *Noah*; or by some ... *National Flood*, such as the *Orygian*, or *Deucalionian* [floods recorded in classical literature][71] than either which there is nothing more improbably'. He argued that Noah's flood had not been universal and was certain that it had not reached western Europe. He followed Ray in arguing that animal remains would have been carried down to the sea by any flood, rather than up mountains, and offered the empirical evidence that in Oxfordshire fossils are rarely found in valleys.

The ultimate difficulty was that, if these fossils represented once-real organisms, then he had to accept the reality of extinction. That was Plot's sticking-point: 'If it be said, that possibly these Species may now be lost, I shall leave it to the Reader to judge, whether it be likely that Providence, which took so much care to secure the Works of the Creation in Noah's Flood, should either then, or since, have been so unmindful of some

Shell-Fish (and of no other Animals) as to suffer any one Species to be lost.'

The Greeks and Romans had wondered about fossils and at least some ancient philosophers had realised that they must represent once-living creatures and that therefore there must have been great changes in the surface of the earth in order for them to be turned to stone and distributed on hills and mountains far from the seas in which they once lived. Leonardo da Vinci, who had thought carefully about almost everything, saw that fossils must be real, and if they were found on mountains, then the sea must once have covered the earth in those places, the fossils having been formed on an old sea bed, the sea since having retreated. But sometimes each generation has to discover things for itself and the idea that fossils were somehow an artefact of the rocks themselves took a long time to die.

In 1671, the influential Oxford scholar Martin Lister presented at the Royal Society a paper on 'Fossil Shells in Several Places of England' in which he argued that fossils were the creations of a natural property of the ground, a 'plastick virtue' that was directly analogous to the properties of 'generation' in living creatures.[72] This seems far-fetched to our eyes, but has a certain logic. A seed growing in the soil produces a whole tree; all life proceeds in some sense from the soil and seems therefore at least partly to be caused by it. Perhaps the earth could also cause imitation creatures in the form of artefactual 'formed stones'.

In the spirit of the new scientific age, however, that extraordinary polymath Robert Hooke had already devoted a great deal of thought to fossils. Some of his ideas were published in his great work on microscopy, *Micrographia* (1665), where his skill in drawing a number of the finest specimens was invaluable in demonstrating to any reader that fossils were not to be confused with curiously shaped stones or some mysterious process within the rocks for which there was no empirical evidence at all. He concluded that 'it seems quite contrary to the infinite Prudence of Nature, which is observable in all its Works and Productions . . . that these prettily shaped Bodies [should have been] generated or wrought by a plastick Vertue, for no higher End than only to exhibit a Form.' In his later *Discourse of Earthquakes*, Hooke stated: '[One should not] imagine that these qualified bodies shall, by an immediate plastick Vertue, be thus shaped by Nature contrary to her general Method of acting in all other Bodies.' He demonstrated quite conclusively that fossils were the remains of real animals, formed either directly by the action of 'some petrifying liquid Substance' or as impressions made directly in sediments from the real objects (or, of course, both) and 'not from the imaginary influences of the Stars, or from any Plastick faculty inherent in the Earth itself.'[73]

Hooke's arguments were so sound, and his brilliant illustrations of fossils so obviously represented nautiloids, shark teeth, mastodon teeth, a crab, crinoid stems and plant remains, that the logic of his presentation seems impossible to refute. Perhaps most telling, after the visual evidence of the fossils themselves, was that one could discern a time-line in their formation. In superficial geological deposits like peat and unconsolidated sands, one finds organic remains that are incompletely fossilised. As one goes to deeper and older versions of the same kinds of deposit, say from 30 million year old Oligocene lignites or the 300-million year old Carboniferous coal measures, the fossils become more and more transformed into rock. Therefore, fossils are formed by natural processes from the remains of real organisms over long periods of time. Hooke also concluded that if fossil-bearing rocks are now found in hills and mountains far above any original sea level, then the earth must have been moved and raised up and the organic remains with them – by the action of earthquakes and volcanoes.

Dr Plot did not give up easily. Where Hooke and others argued that the resemblance between fossils and their living models was too great to explain away, Plot replied that it was all coincidence and, in any case, many formed stones were not so realistic. And even if some actually were the remains of real shellfish, it does not follow that all were. Hooke had even pointed out that some fossils still had the original shelly material, and even colour, preserved. But Plot had a ready answer for that: those were real shells that had contaminated the fossil samples, as, for example, in the great Roman oyster midden near Reading. Furthermore, Plot asked, if fossils were real, why did one not find more bones of whales or the shells of crabs and lobsters?

Hooke put the question in reverse: 'If these be Apish *Tricks of Nature*, why does it not imitate several other of its own works?' To this Plot answered, piously, that 'Nature herein acts neither contrary to her own *Prudence, human Ratiocination*, or in *vain*, it being the Wisdom and goodness of the *Supreme Nature* . . . to beautify the World with these Varieties.' In this, nature acted as it did with respect to 'most *Flowers*, such as *Tulips, Anemones* &c. of which we know as little Use as of *Formed Stones*'. Among other technical objections, Plot noted that some formed stones are found in beds close together, and others scattered, while some 'represent things of so tender a Texture, and of so short a Duration, that it is very improbable . . . [they] should ever continue long enough [to be formed into fossils]'. Plot's weakest argument (but still a popular way of debating today) was to say that because a fossil ammonite was not perfectly identical with a modern nautilus, it could not be 'real'.

Plot held out the additional hope that an explanation of formed stones might be found in the 'spontaneous Inclinations of Salts', that is, in the processes underlying the structure and formation of crystals, and it has to be admitted that there was much common sense in that. Many a tyro fossil collector, this author included, has been fooled into thinking he or she has found a beautiful fossil fern, only to discover it is a branching crystal structure formed by the penetration of mineral-laden water into the bedding plane.

Hooke was one of Plot's closest scientific friends and contemporaries at Oxford, as was Martin Lister. Plot tried to incorporate the ideas of both friends in his book, but Hooke and Lister held diametrically opposing views. All this contributed to Plot's dilemma. Everything came down to a single question:

> Whether the Stones we find in the Forms of Shell-fish, be *lapides sui generis* [stones that have formed themselves], naturally produced by some *extraordinary plastic virtue*, latent in the Earth or Quarries in which they are found? Or, whether they rather owe their Form and Figuration to the Shells of the Fishes they represent, brought to the places whether they are now found by a Deluge, Earth-quake, or some such means, and there being filled with *Mud, Clay*, and petrifying *Juices*, have in tract of time been turned into Stones, as we now find them, still retaining the same Shape in the whole, with the same *Lineations, Eminencies, Cavities, Orifices, Points*, that they had whilst they were Shells?

He concluded, reluctantly: 'Upon mature Deliberation, I must confess I am inclined rather to the Opinion of Mr *Lister*, that they are *Lapides sui generis*.' He rejected the view of 'those Eminent *Virtuosi*, Mr Hook and Mr Ray ... *that they are ... formed in an Animal Mould*. The *latter Opinion* appearing at present to be pressed with far more, and more insuperable Difficulties than the *former*.' Then, having done his best to set out the arguments on both sides, Plot, with all modesty and caution, qualified *his conclusion*:

My present opinion, which has not been taken up out of Humor or Contradiction, with intent only to affront other worthy Authors modest Conjectures, but rather friendly to excite them, or any others, to endeavour collections of shell-fish and parts of other Animals, that may answer such Formed Stones as are here already, or may hereafter be produced: which when I find done, and the reasons alleged solidly answer, I shall be ready with Acknowledgement to retract my *Opinion*, which I am not so in love with, but for the sake of *Truth* I can cheerfully cast off without the least Reluctance.

One such who made an extensive collection of fossils was Edward Lhwyd (?1660–1709), a poor student at Oxford whom Plot took under his wing. Lhwyd's greatest achievement came in 1698 when he discovered trilobites, thinking them 'doubtless [to] be referred to the skeleton of some Flat-Fish'. Trilobites would later turn out to be critical in dating the earlier stages of the fossil record.[74] Lhwyd assisted Lister in cataloguing the fossils of Oxfordshire and in 1699 published his *Lithophylacii Botannici Ichnographia*,[75] representing his life's work on fossils. Caught like Plot between two worlds, old and new, Lhwyd ingeniously explained fossils as the result of a 'Spermatick Principle' in the rocks – essentially the growth within the rocks, under the influence of water, of tiny seeds that had been placed there at creation. The strength of Lhwyd's theory was that it did not require the Flood or earthquakes to strew them about. He summarised his ideas and arguments, which owe something to St Augustine of Hippo, in a long rambling letter to John Ray, published in the later editions of Ray's *Three Philosophical Discourses*.

Where Plot had a running intellectual argument with Hooke and Lister, John Ray similarly contended with Lhwyd, coming down firmly in favour of the view that 'formed Stones, Seashells, and other Marine-like Bodies found at great Distance from the Shores, supposed to have been brought in by the Deluge' had all been formed in situ from once-living creatures. After very sympathetically considering Plot's and Lhwyd's theories, he concluded: 'these Bodies owe their Original to the Sea, and were sometimes the Shells or Bones of Fishes.' That being the case,

Edward Lhwyd

the distribution of fossils must be due to forces that had changed the whole face and structure of the earth. Ray rejected the Flood as the cause of such perturbations and explained them instead, like Hooke, in terms of earthquakes and volcanoes (all of which acted to create a better world for man to inhabit). But Ray could not go so far as to accept the reality of extinction, 'as there neither is nor can be any new Species of Animals produced ... so Providence ... doth ... watch over all that are created, that an entire Species shall not be lost or destroyed by any Accident'. The only possible conclusion, therefore, was that they must all be living somewhere still, just as 'Wolves and Bevers, which we are well assured were sometimes native of England [yet there remain] Plenty of them still in other Countries'.

These debates, vacillations and compromises capture the spirit of the time. As far as fossils were concerned, matters were at something of a stand-off. If it had been possible to agree upon a mechanism for the raising up and tearing down of mountains, it would have been easier to accept 'formed stones' as the remains of once-living organisms. Equally, if there had been conclusive evidence that fossils were real, then natural philosophers would have had no alternative but to accept the fact of the changing structure of the earth. As with most ventures in life, the work proceeded iteratively, stumbling from incomplete data to premature conclusion and back again. But in the end, the reality of fossils and the fact of a changing earth prevailed and, as this happened, another common phenomenon came into play. Throughout history, we find that an idea in isolation may not be dangerous or controversial until it becomes associated with other ideas. Fossils might have been safe on their own, but when the Church was under threat from all sides, fossils became dangerous. Similarly, as we will see later, ideas about evolution were tolerable or dismissible in the eighteenth century but anathema in the nineteenth. And any idea becomes more dangerous when it is backed up by fact. Fossils by themselves were interesting natural phenomena; fossils of extinct creatures, in an ancient changing earth, contradicted the Bible. In the context of other philosophies and theories of organic change, they challenged the very notion of the Creator.

Meanwhile, Plot had saved a special puzzle for the later stages of his long discussion. He illustrated a fossil from the village of Cornwell, in the western part of Oxfordshire. Just the sort of 'stone' that Paley could have pitched his foot against, it is so clearly part of a bone – in fact the distal end of a femur (thigh bone) of a large vertebrate – that its organic nature and origins cannot be denied. Plot saw that it could be nothing but 'the lowermost part of a *Thigh-Bone* of a *Man*, or at least of some other *Animal*, with the *capita Femoris inferiora*, between which [are] the *anterior* . . . and the *posterior Sinus*, the seat of the strong *Ligament* that rises out of the Leg.' This fossil bone, fifteen inches long and weighing twenty pounds, must have come from a large animal – very large, in fact.

Plot's wriggling on the hook of this conundrum is almost painful to read. Firstly, it 'must have been a real *Bone*, now *Petrified*, and therefore, indeed not properly belonging to this place'. But against this, it was larger than any known thigh bone of any animal, 'both *Horses* and *Oxen* falling short of it'. Perhaps, then, the process of petrification had caused it to be enlarged – but he rejected that idea. His next solution was that 'in probability [it] must have been the Bone of some *Elephant*, brought hither during the Government by the *Romans* in *Britain*'. But he knew of no record of Hannibal or anyone else bringing elephants as far as Britain (actually, Claudius did) and there was no sign of any tusks from the elephant. In fact, an even bigger bone, similar to Plot's, was reported to have been found when St Mary Wood Church was pulled down after the Fire of London in 1666, and another was reported from the grounds of a church in Gloucester. Plot toyed with the idea that the Emperor Claudius was involved after all and then, just at the moment he was writing his *Natural History*, a living elephant was put on show at Oxford and he was able to compare its bones and teeth with his fossils. He found serious differences, so Roman elephants were ruled out.

This left only one possible answer: 'It remains, that (notwithstanding their extravagant magnitude) they must have been the Bones of *Men* or *Women*.' And so Plot retreated to the Bible, where there is confirmation that, in the words of Genesis 6:4, 'There were giants in the world in those days.' He noted that Goliath had been 'nine Foot nine Inches' tall and quoted other cases of giants from the classics, from fables, and closer to home. 'The tallest that I have seen in our Days, was ... a woman of Dutch Extraction, shewn publickly here at Oxford, seven Foot and a half high, with all her Limbs proportionable.' Thus, Plot's huge bone must have been nothing more, or less, than a remnant of an ancient giant from the days of antiquity. A most convenient answer but as unsatisfactory to Plot, one suspects, as it is to us. Once again we have seen the application of the dangerous mode of argument (associated with the name Sherlock Holmes) that when you have eliminated the possible, what remains, however improbable, must be the truth.

We now know that Plot's 'thigh bone of a giant' is something scarcely less plausible – it is the tip of a femur from a nine-metre-long extinct 'lizard', a Late Jurassic dinosaur 160 million years old. In 1824, William Buckland at Oxford described a collection of similar bones from Stonesfield in Oxfordshire and showed that they indeed came from a giant – but a giant lizard-like animal standing four or five metres high. Its huge teeth show it to have been a carnivore, a giant predator of the Jurassic plains (and heaths). Plot had given the first ever description of a *dinosaur* fossil. Buckland gave it the name 'megalosaurus'. In 1841, when two other kinds of similar behemoths had been found – iguanodon and scelidosaurus – Richard Owen in London created the new taxonomic category 'Dinosauria' ('terrible lizards') to include them. Buckland's megalosaurus is still on view in the Oxford University Museum, but Plot's specimen, to Oxford's eternal regret, has been lost or stolen.[76]

By the time Paley came to write his version of natural theology, the true nature of fossils was more or less generally accepted.

Extinction was also an accepted fact in many people's minds
– but not in all. The alternative was Ray's proposition; those
animals and plants that seemed to have died out might simply
be lurking in some distant unexplored place. This, at least, was
a testable idea. In 1796 William Clark (later of Lewis and Clark
fame) collected for Thomas Jefferson the remains of extinct
mastodons from Big Bone Lick in Kentucky (just across the
Ohio River from modern Cincinnati).[77] We now think that
mammoths and mastodons might in fact have been driven to
extinction, or at least their last dwindling representations killed
off, by humans. Jefferson, however, was unwilling to accept the
concept of extinct animals. One of the non-political aims of
the Lewis and Clark expedition was to search for the relict
populations of mastodons in the far American West. Every now
and again the idea resurfaces, for example in the myth of the
Loch Ness plesiosaur, and is of course a staple of fiction such
as Conan Doyle's classic *Lost World*.

The progress of palaeontology and the underlying geological
science was much slower in Europe than it would have been if
the early geologists and palaeontologists had been living in the
American West or Australia, where a relative absence of vegetation not only accelerates the processes of erosion (all those
canyons and 'badlands' where dinosaurs seem to stick up out
of the ground) but simply makes the structure of the earth easier
to see. In Europe, except on mountains above the tree line, most
of the earth is covered under grass, trees and bushes. Not until
the seventeenth century, when men started to hack deep into
the earth to look for minerals, to mine coal, iron ore and limestone, and to build roads, canals and railways, did something
of the underlying structure start to emerge.[78]

By Charles Darwin's day, fossils were as fully embedded in
theories of a changing earth as they were literally embedded in
that earth. While many serious philosophers contributed to that

understanding, a good deal of the credit for teaching us how to read the story of fossils must go to the great self-taught English geologist William Smith (1768–1839). Because of his hands-on experience as a canal and mine surveyor, Smith was able to tease apart the structure of the earth in ways that no theorist could. He started as a surveyor of mines and canal excavations and ended (not without controversy) as the founder of stratigraphy. This was empirical science on the scale of hundreds of miles, not laboratory experiments or nature observed under a hand-lens. Smith discovered the patterns underlying the consistencies and inconsistencies in the surface of the earth. One of his key observations was that unique fossils and assemblages of fossils were consistently associated with particular kinds of rocks. The fossils were a signature that allowed him to fit outcrops of rock, small or large, separated by hundreds of miles, into a pattern. On his exploratory rides across country and wherever his workmen dug away the overlaying vegetation and silt and cut into the bedrock, more details were added.

Smith's genius was to see how they all fitted together; he started to trace the lateral extent of individual beds right across landscapes and then read the structure of the earth by virtue of joining together the dots – the places where digs had been made. He saw that, where strata had become tilted, a canal excavation would run at an angle to the tilt, exposing layer after layer lying on top of each other. Thus he could also create a three-dimensional view of the structure of the earth with all its folds and faults. Smith then published his results in the form of the first geological maps. Naturally these maps had huge value for the mining and canal building industries: 'Here will be coal, under there will be iron ore' and so on. So, of course, others tried to claim the credit and to publish rival maps and take the financial benefit. A better geologist than businessman, Smith was forced into bankruptcy, but in the end his reputation was recovered. Smith's maps delineating the geological history of the earth pointed the way to a new set of truths. Geology had moved from a local to a regional, national and even global scale.

CHAPTER SEVEN

Sacred Theories

'All mountains at this day have not existed from the beginnings of time . . . Yet this is certain, that a great parcel of the Earth is every year carried into the Sea [and] does leave new lands fit for new inhabitants.'

Steno, *Prodromus*, 1669

'A World lying in its own rubbish . . .'

Thomas Burnet, *Telluria Theoria Sacra*, 1681

'I have been working at so many things: that I have not got on much with Geology: I suspect the first expedition I take, clinometer & hammer in hand, will send me back very little wiser & [a] good deal more puzzled than when I started. – As yet I have only indulged in hypotheses; but they are such powerful ones that, I suppose, if they were put into action but for one day, the world would come to an end.'

Charles Darwin, letter to the Reverend John Stevens
Henslow, 11 July 1831

If Charles Darwin was strongly influenced by reading Humboldt, Herschel and Paley in 1831, a book he was given later that same year turned out to be even more important. Although we customarily think of Charles Darwin as the naturalist par excellence (a student of animals and plants, their behaviour and their ecology), he actually began his career intending to be a geologist. During the August preceding the voyage on HMS *Beagle*, he had spent time in Wales as assistant to Adam Sedgwick, Woodwardian Professor of Geology at Cambridge, where he evidently had sufficient skill to make original, if modest, contributions to Sedgwick's unravelling of the structure of the Palaeogoic strata in that region. His excitement about geology shows in the letter he sent to his mentor Henslow, quoted above. Later, while circumnavigating the world on HMS *Beagle*, he planned what he thought would be his first major work, a geology of South America analysed in thoroughly modern terms: a geology of change.[79]

When he left England on the *Beagle* in late December 1831, Darwin carried with him a present, a copy of the brand-new first volume of Charles Lyell's *Principles of Geology*.[80] Through this work Lyell intended to make geology the next great subject to become fully scientifically founded, after cosmology and mechanics (chemistry would follow, and biology turned out to be last). Lyell's view of the changing earth was captured in the picture he used as a frontispiece, of the marble pillars of the Temple of Serapis at Puzuoli near Naples. Just one piece of stone tells the story of millions of years. The marble was originally formed as a limestone sediment at the bottom of the sea, buried and metamorphosed into marble, and (millions of years later) raised up as dry land. Some two thousand years ago, the marble was quarried, carved into pillars and set into the building. But the marble shows clear evidence that – after it was carved and set into place – it had been 'pierced by a species of marine perforating bivalve [a clam relative] . . . [and] at the bottom of the cavities, many shells are still found, notwithstanding the great numbers that have been taken out by visitors'. Therefore, the earth has not been still; sometime after it had been carved the whole structure had been partially submerged under water. Now it has been raised above water yet again.

One might expect that Darwin had been given the copy of Lyell by his Cambridge teacher, the Reverend John Stevens Henslow. But Henslow very much disapproved of Lyell's revolutionary new geology, warning Darwin 'on no account to accept the view therein advocated'. Neither was it given to Darwin by Sedgwick, who was no fan of Lyell either. In fact, it was a gift from Robert FitzRoy, captain of the *Beagle* and one of the most scientifically inclined officers in the Navy. Ironic, then, that this book should have been the one to set Darwin on his scientific career while thirty years later FitzRoy, having newly discovered a fundamentalist religion after the voyage, should have become a bitter foe of his old shipmate and his ideas.

Once the *Beagle* set sail, Darwin discovered seasickness, an affliction so systemic that it never left him, even after five years at sea. He retired to his hammock after the first day and stayed there. Soon he began to read Lyell. When they made their first landfall at Sān Tiago in mid-Atlantic, he found that he could easily read in the cliffs exactly the kind of sequence that Lyell had demonstrated at the Temple (actually a market) of Serapis. Like so many others, Darwin realised that a whole world of scholarship was wrapped up in the 'simple' stones and rocks, and he resolved to become a serious geologist. Eventually, while the study of living creatures led scholars of the nineteenth century to dare to ask the simple question, Are species always fixed? – and led to Darwin's unravelling of the puzzle of 'origins of species' – it was the study of geology that had blazed the trail.

But Darwin was only a very small part of a very long tradition of geological study, much of which, rather than supporting evolution, was designed to support the biblical narrative in Genesis. Even when John Ray was writing *The Wisdom of God*, a great deal of geological scholarship was considerably at variance not only with the Bible but also with common sense. The earth does not appear to be changing beneath our feet. The mountains and 'eternal hills' seem permanent. The oceans, for all their storms and tides, are constant and predictable. The common-sense view surely would be that the earth is static and unchanging, reliable as the phases of the moon, constant as the sun itself – (until Galileo had inconveniently – and heretically – discovered its flaws). The Bible tells that this world is not the end, but will itself be consumed in a great conflagration. But it will remain static and unchanging in the interim, as common sense (and Aristotle) suggested. At least that seemed to be what the Bible meant when God said: 'While the earth remaineth, seed time and harvest, and cold and hot, and summer and winter, and day and night shall not cease' (Genesis 8:22). The issue perhaps lies in the words 'while the earth remaineth', which could be interpreted either to allow that natural processes will change it, or that God simply reserves the right to destroy it as often as he sees fit.

In the late seventeenth century, as fast as new geological ideas developed, so equally it became essential that these discoveries and insights be enfolded in religious orthodoxy, and particularly that scholars should devise ways to reconcile them with the Bible. Worried clerics and philosophers had to find ways to co-opt it, not as a cynical exercise in damage control but a manifestation of the belief that rational enquiry ought to lead to new revelations of the power and mysterious ways of God. The real danger was that the Church would not have the intellectual fire power to keep up in the race for discovery and interpretation. The result was a whole series of works reconciling, explaining, extrapolating, interpreting, reinterpreting (every now and again misinterpreting) and combining science and biblical authority into one uneasy system. Perhaps the most influential were those by the Reverend Thomas Burnet, the Reverend John Ray, and his disciple William Derham, but there were also the Reverend Doctors Ralph Cudworth, Samuel Clarke, John Woodward, William Whiston, Richard Bentley, Philip Howard, and countless others. Many of these works were first delivered as Boyle Lectures. Reading them today, we can only admire the brilliance with which they tried to make both scientific and theological sense out of incomplete and unsatisfactory data. Whether they seem to have been ahead of or behind their times, they are part of our history. They were all, like Gilbert White or Robert Plot, John Ray or even William Paley, caught on the intellectual knife-edge between the old and the new, between materialist science and religious orthodoxy.

One of the first great attempts to bring together in one unifying theory the Biblical account of creation and the new sciences of the earth was the brainchild of Thomas Burnet, yet another outspoken and eloquent Cambridge-educated free thinker of the Newtonian era. He was educated at Clare Hall but, as a follower of Ralph Cudworth, became a Fellow of Christ's College in 1654 and later travelled in Europe with Lord Wiltshire and Lord Orrery, grandson of the Duke of Ormonde. Through the influence of the latter, he was appointed in 1687 to a position of significant influence as Master of Charterhouse School. At a time of much mutual suspicion between Protestants and Catholics, he had a chequered career in the Church, eventually becoming Chaplain-in-Ordinary to King William III. Despite his powerful connections, he rose no further in the ecclesiastical hierarchy because his scholarly ideas on the nature and history of the earth that first brought him some fame, then, because they conflicted with the biblical account of creation, created considerable notoriety. His *Telluria Theoria Sacra*, (or *A Sacred Theory of the Origin and general Changes of our Earth* of 1681[81]) is notable both for the sweep of its thought and subject matter and for its grand, lyrical prose. *Theoria Sacra* is a combination of Ovid's *Metamorphoses* and Ussher's biblical literalism, with shades of Blakean mysticism. But it is the Cartesian influence that most characterises this extraordinary book, which more or less started physico-theology in Britain.

In his *Principles of Philosophy* (1644) Descartes had based his cosmological theories on the premise that a spherical, rotating body like the earth or sun must once have been a hot, chaotic fluid of atoms. The earth, flung into its orbit around the sun, gradually cooled and its component parts sorted themselves out into five shells or layers: an inner unknowable core; an interior crust of metals; then water; then air; the surface crust as we know it; and finally the air above (the atmosphere). This account of the sorting of the original 'elements' of the earth into regions depended in part on scientific principles like specific gravity and partly, apparently, on the authority of Genesis (hence the air and water under the earth's outer solid crust). Next, the outer crust collapsed inwards, forming the physiographic features of the present earth including the mountains and, by releasing the abyss, the oceans. At this point, our familiar earth, with its phenomena of earthquakes, volcanoes, rivers, lakes, clouds and rain, began.

Here then is what looks awfully like an atheistic philosophy in full flow: atoms, chance, chaos, too much time, an ancient earth, too many lawful processes. Nowhere can we see the hand of God except, grudgingly – or tactfully, perhaps – in the very moment of origin itself. Burnet set out to make this materialist cosmology more or less compatible with the very words of Genesis. He opened his sacred theory, however, on very safe grounds with a diatribe against Aristotle. This might seem odd, but it was part of a complex strategy. Burnet had to create as orthodox a theory as possible in order to gain acceptance for his big heretical idea, which was the very modern notion that instead of remaining unchanged since creation (except for the effects of the Flood), the earth was actually in flux and subject

Thomas Burnet

to powerful, continually acting forces. As Burnet could scarcely challenge the Bible head-on, he chose everyone's favourite lateral target, Aristotle, who, although the one true source of non-biblical authority in medieval times, now stood for the Dark Ages. He had taught that the earth was eternal, that it had existed since eternity as it was now, unchanged and unchanging, and would so continue: an easy target if the Bible was right and accurately described a beginning (dated by Ussher) and forecast a most definitive end.

So Burnet began his thesis by listing all the evidence he could find that the earth shows signs of change: everything from the numbers of people, to earthquakes, the rise of industry and inventions, exploration, and 'use of the Loadstone'. This then gave him the opportunity to fire an opening salvo about the earth itself. The earth could not be eternal, precisely because it is constantly changing as a result of observable processes. That it had not been changing for very long was self-evident: '[If] this present state and form of the Earth had been from Eternity, it would long ere this destroy'd it self, and chang'd it self: the Mountains sinking by degrees into the Valleys, and into the Seas, and the Waters rising above the Earth.' This clever argument neatly killed two birds with one stone. He had introduced his scientific premise that the earth was changing and had protected his rear by insisting that it could not have been changing for very long (his religious premise). Now he was positioned to make the case for that changing earth.

> If I was to describe [the earth] as an Oratour, I would suppose it a beautiful and regular Globe, and not only so, but that the whole Universe was made for its sake; that it was the darling and favourite of Heaven, that the Sun shin'd only to give it light, to ripen its Fruit, and make fresh its Flowers; and that the great Concave of the Firmament, and all the Stars in their several Orbs, were ere design'd only for a spangled cabinet to keep this jewel in.
>
> But a philosopher that overheard me, would either think me in jest, or very injudicious . . . this, he would say, is to make the great World like one of the heathen temples, a beautiful and magnificent structure, and of the richest materials, yet built only for a brute Idol, a Dog, or a Crocodile, or some deform'd Creature, plac'd in a corner of it. We must therefore be impartial where the Truth requires it, and describe the Earth as it is really in its self; . . .'tis a broken and confus'd heap of bodies, plac'd in no order to one another, nor with any correspondency or regularity of parts: And such a body as the Moon appears to us, when 'tis look'd upon with a good Glass, rude and ragged . . . a World lying in its rubbish.

A 'broken and confus'd heap of bodies . . . lying in its own rubbish': Burnet reached these conclusions after a trip to the

145

Alps, being neither the first nor (certainly) the last to find poetic or scientific inspiration in these great jagged mountains, so awesome compared with tamer British landscapes. 'These Alps we are speaking of are the greatest range of Mountains in *Europe*; and 'tis prodigious to view and consider of what extent these heaps of Stones and Rubbish are . . . a multitude of vast Bodies thrown together in confusion. As those Mountains are . . . for those that live amongst the Alps and the great Mountains, think that all the earth is . . . all broken into Mountains, and Valleys, and Precipices.' In fact, Burnet saw evidence of catastrophic change everywhere. Once he had seen it in the Alps, he realised that it even underlay the green meadows of England.

In this view of a shattered, ruined earth, with its echoes in Hume's view of nature's flaws, Burnet was unusually modern and bold. It takes a great leap of imagination to feel Burnet's shattering of the earth and to sense the precarious grip that humans have on it, especially if you have never seen the Alps. As a Christian, Burnet needed not only to explain such devastation, he needed a philosophy and a natural history to reconcile it with God's plan and God's purpose. The result was his unique sacred history of the earth (with bits cheerfully borrowed from Descartes), accounting in geological terms for the major milestones: Creation, the Flood, the Present Earth, and the Final Conflagration and accompanying Redemption according to the Book of Revelation.

In Burnet's theory, carefully crafted to accommodate the account of Genesis, there were four early stages in the earth's history. Any sacred theory must first account for the beginning, when 'this earth was without form, and void; and darkness was upon the face of the deep. And the Spirit of God moved upon the face of the waters'. Traditionally, the words have been taken to mean that the first form of the earth – after nothing at all – was a uniform and unstructured ocean ('the deep') and that

God created everything from that starting point. This, however, could scarcely satisfy scholars of the seventeenth century who, after Descartes, already had concluded that the earth must have had a fiery origin like the sun, or even from the sun. The 'void' therefore had to be reconstructed as a 'chaos' of unformed matter, just as Epicurus had proposed. First there was this chaos and then God created order. (At this time, it was common that the words 'In the beginning' should be construed to signify an indeterminate period of time, prior to the six days of Creation.)

Next, God imposed order on the chaos. The obvious mechanism that would allow the ordering of the 'stuff' of the world on fully scientific principles was specific gravity (this being the hot subject of the day[82]). The result was that the elements making up the chaos (fire, water and air) became arranged by weight, producing a fiery inside, with an outer layer of water, all surrounded by the air and the 'firmament' or heavens. This second stage therefore produced a completely aqueous surface for the earth and explains the 'deep' of Genesis, verse two. Meanwhile, the inside cavity of the earth had now become a curious mixture of inner seas and subterranean caverns and immense fires, a truly Blakean world. The 'subterraneous cavities' were the source of '[volcanoes] or fiery Mountains; that belch out flames and smoke and ashes, and sometimes great stones and broken Rocks, and lumps of Earth, or some metallick mixture ... these argue great vacuities in the bowels of the Earth.'

Formation of the earth's crust followed. It is implicit in the ninth verse of Genesis that there must have been solid matter under the oceans. God apparently had simply divided the water from the dry land. Having postulated a fiery centre of the earth deep underneath the oceans, Burnet needed to find a different origin for the earth's crust and here he is at his most ingenious. First he assumed that the original material of the chaos – including 'materials and ingredients of all bodies' – was not all deposited in the centre of the earth, but that some was left dispersed in the air. Meanwhile, the oily parts of the original chaos would have risen to the surface of the waters. Sediment from the 'thick, gross, and dark' air gradually settled on the oily surface like so much snow, where it became trapped, 'mixing there with that unctious substance ... [where they] compos'd a certain slime, or fat, soft and light Earth, spread upon the face of the Waters'. Eventually the solid surface of the earth built up 'smooth, regular and uniform, without Mountains, and without a Sea' (the seas now being underneath the land).

So far, Burnet's earth was smooth and pristine, but eminently suitable for animals and plants (except possibly creatures like mountain goats). In fact, life on this earth was luxuriant. Burnet needed next to explain the present 'broken' configuration of the earth, which must have required universal and overwhelming changes. That was simple: the great Flood.

The Flood might almost be ignored as no more than a brutal story if it were not for the fact that a flood legend appears in the folk history of cultures from America to China. Two accounts of it appear in the Bible.[83] But if it really happened, it must have left a mark on the face of the earth that, if we could find and properly interpret the evidence, would provide direct vindication (indeed a proof) of the authenticity of the Bible story. Philosophers had long argued about the source of all the water for the biblical flood. The volume required to cover all the mountains would have been far more than rain alone could have produced in the space of forty days and nights. Here again, Burnet appealed to Genesis. Where the Bible says that 'All the fountains of the deep were broken up and the windows of heaven were opened' (Genesis 7:11), Burnet saw an explanation of his 'ruin'd earth' – the very breaking open of the earth, like the smashing of an egg, to release the fountains of the deep was the same catastrophe that threw up our present mountains, created deep valleys and gorges, overhanging precipices, folded strata and the geological evidence of change and destruction. Heat and pressure had built up under the pristine surface of the earth and it had all burst assunder.

No one before Burnet had attributed such a mass deconstruction of the earth to the Flood. There is no account in the Bible to corroborate it. When Noah and his family were released from the ark, there was no indication that the earth (though no doubt a lot wetter and muddier) was substantially changed from before, let alone riven apart. The animals disembarked to familiar territory in which plants and trees sprang up again. The Bible said, 'behold, the face of the ground was dry'.

The reason for all the petrological mayhem was obvious. The shattered and incoherent state of the earth is a punishment for man's sin, a sign of God's anger with man for his imperfections and excesses. 'The earth was also corrupt before God, and the earth filled with violence. And God looked upon the earth and, behold, it was corrupt . . . and God said to Noah the end of all flesh is come before us . . . I will destroy them with the earth.' (Genesis 6:12–13). Thereafter, all would slowly disintegrate from erosion, earthquake and volcanoes, man having lost the right to live in the perfect earth that God originally created. The earth has been, and will continue to be, heir to a thousand minute, petty insults. Erosion will slowly wear away the land and earthquakes and volcanoes will break it further apart, 'the Mountains sinking by degrees into the Valleys, and into the Sea . . . for 'tis certain, that the Mountains and higher parts of the Earth grow lesser and lesser from Age to Age, and that from many causes, sometimes the roots of them are weakened and eaten by Subterranean Fires, and sometimes they are torn and tumbled down by Earthquakes. Winds, Rains, and Storms,

149

and the heat of the Sun without [consume them insensibly]'. This period of constant change and decay will only end when the earth is consumed in a final conflagration at Judgement Day.

This, then, is what became known as a physico-theological argument. And his argument cuts some corners. Because Burnet explained the origins of mountains from the single cataclysm of the Flood, it followed that no further mountain-building would occur. He dismissed out of hand those contemporary geologists who said that 'Mountains, and all other irregularities in the Earth, have risen from Earthquakes, and such causes'. Burnet asserted that 'Earthquakes seldom make Mountains, they often take them away and sink them down into the Caverns that lie under them ... Who ever heard of a new chain of Mountains being made upon the Earth, or a new Channel made for the Ocean?' And he adds one telling point, a difficulty that would be problematic for the next 300 years: 'Besides, Earthquakes are not in all Countries and Climates as Mountains are.'

On the other hand, all that remains after the flood is subject to inexorable decay,

> for whatever moulders or is washt away from them [the mountains], is carried down into the lower grounds, and into the Sea, and nothing is ever brought back again by any circulation; Their Losses are not repaired ... The air alone, and the little drops of rain have defac'd the strongest and proudest monuments of the Greeks and Romans; and allow them but time enough and they will of themselves beat down the Rocks into the Sea, and the Hills into the Valleys.

Burnet's view of the present earth is therefore one of progressive destruction, a linear process, as is essential to his theology of the end to come. But he was willing to allow that the processes of change in the earth might be lengthy: 'I do not say the Earth would be reduced to the uninhabitable form in ten thousand years, though I believe it would but take twenty, if you please, take an hundred thousand, take a million . . .' So time is allowed prospectively, but not retrospectively, and not infinitely, because all is being prepared for the final chapter.

Finally, the end will come sometime when the earth has been made completely smooth again and covered in water. The subterranean caverns will open up a second time and spew fire over the whole earth, which will be consumed in the subsequent conflation according to prophesy. 'Surely in that day there shall be a great shaking in the land of Israel . . . and mountains shall be thrown down, and the sharp places shall fall . . . and I will rain . . . an overflowing rain, and great hail stones, fire, and brimstone' (Ezekiel 38:19–22).

Burnet's sacred theory was exciting, dynamic and dramatic. For all its faults, it tried to make study of the earth compatible with biblical authority. But it was dangerously close to blasphemy in places. It contradicted the Bible and invoked Decartes too often. He had all the mountains being created at the Flood, but the Bible states clearly that they existed before then (Genesis 7: 19–20). He saw a ruined earth and one that constantly changes, directly contradicting God who had said (after the Flood): 'I will not again curse the ground any more for man's sake' (Genesis 8:21).

Burnet had created a great enigma and a quandary. His sacred theory struck a little too close to some cherished beliefs; but the underlying science was not quite good enough to overcome the reservations of orthodoxy. A contemporary satirical ballad had Burnet believing 'That all the books of Moses / were nothing but supposes'. In the end, he was seen as not far from the heresy of which Descartes stood accused – that his philosophy didn't need God at all. Burnet did not help the cause in his later

Archaeologiciae Philosophiae (1692), which included a rather fanciful conversation between Eve and the serpent in the Garden of Eden. The satirist enjoyed this one too: 'That as for Father Adam / and Mrs Eve, his Madame / and what the Devil spoke, sir / were nothing but a joke, Sir.'

Nonetheless, one has to have more than a grudging admiration for the ingenuity of his theory and its cleverness in fitting Descartes' cosmology on the one hand and biblical narrative on the other into a perfectly orthodox framework of creation to final conflagration. The great Isaac Newton wrote to Burnet approving his attempt to confirm the truth of Genesis by science without getting hung up on the literal word: 'I do not think [Moses'] description of the creation [in Genesis] either philosophical or feigned, but that he described realities in a language artificially adapted to the sense of the vulgar . . . Of our present sea, rocks, mountains &c., I think you have given the most plausible account.'[84]

But Burnet had left one major difficulty unsolved. While his theory accounted for the shattered state of the earth, it did not account for the internal structure of the parts that were so shattered. As we pass a major road-cutting we see the layers of rock lying one on top of another. Burnet's Flood might account for the folding and confusion of the major layers and the gaps among them, but not the regular patterning of the different sub-strata within each section. His account failed to explain what had caused the earth's crust, before the Flood, to be so orderly and patterned.

Burnet's theory, crafted in a general spirit of broad rational enquiry, and doubtless with the hope that it would make him very famous, caused a storm of discussion. A howl of anger arose from everyone (and there were many) whose particular oxen Burnet had gored. He had offended the deists with his science and the scientists with his religion. Nonetheless, Burnet had performed a valuable service. After him, it was easier to open up the Bible to scientific interpretations. Burnet was not a geologist, he could only theorise from what he saw. For the next hundred years, progress in geological thinking came in fits and starts. As with Plot's treatment of fossils, reading the works of the period shows how deeply torn many scholars were between what they were learning about the earth and how it might affect their beliefs. If their writings sometimes seem awkward and contrived, we must remember that these were deeply committed scholars trying to make a compromise work.

Fifteen years later, Dr John Woodward (1665–1728) answered Burnet. Woodward was a man whose name, although he never worked or studied there, became closely associated with the University of Cambridge. Something of a mixture between a physico-theologian of the old school and a more modern geologist, he came from a poor but respectable family and was first apprenticed to a linen draper. Evidently his intelligence and ambition soon stood out because at the age of nineteen he came to the notice of the Queen's Physician, Peter Barwick, who took him on as a pupil. Four years later, in 1688, Woodward was sufficiently qualified to work as an independent physician in London, and in 1692 he became Professor of Physic at Gresham College, London. By now he had the means and leisure to exercise his mind with scientific matters like fossil collecting and geology. His *Essay towards a Natural History of the Earth* (1695) made his reputation as a scholar.[85] He may have been no better a scientist than doctor but on his death he left enough money to the University of Cambridge to found the Woodwardian Chair of Geology (to be occupied by, among others, Darwin's geological mentor, Adam Sedgwick).

A conservative by nature, where Burnet saw revolution and the chaotic ruination of a once perfect world, Woodward in his *Essay* tried to find the glimmerings of a law-like order. Where Burnet saw linear change, Woodward saw stability. Wherever Burnet had made a mistake, Woodward leapt in. For example, Burnet had proposed that the earth was originally flat but Woodward pointed out that Genesis refers to mountains existing before the Flood: '[The] Form of the earth, before the Deluge, was not *smooth*, eaven, and uniform: but unequal, and distinguished with *Mountains*, *Valleys* and *Plains*: as also with *Sea*, *Lakes* and *Rivers*.' Therefore the Flood could not have caused the formation of the first mountains and other major topographic features. Woodward could offer no scientific explanation of their formation; indeed, he thought that none was needed. God had simply made the best of all worlds, and if that included mountains, so be it.

Woodward presented a theory about the earth's structure that once again invoked the Flood as the source of energy, but used it in an even more ingenious and 'scientific' way. Where Burnet saw it as smashing up the earth, Woodward concluded that it had *dissolved* the earth.

> During the Time of the Deluge . . . all the Stone and marble . . . all the Metals, all Mineral Concretions, and, in a word, all Fossils whatsoever . . . were totally dissolved and their constituent Corpuscles all disjoined . . . as also Animal Bodies, and Parts of Animal Bones, Teeth, Shells; Vegetables, and Parts of Vegetables, Trees, Shrubs, Herbs; and to be short, all Bodies whatsoever, that were either upon the Earth, or that constituted the Mass of it . . . were assumed up promiscuously into the Water and . . . made up [into] one common confused Mass.

When the force of the Flood subsided, 'at Length, all the Mass . . . was again precipitated . . . as nearly as possibly could be expected in so great a Confusion, according to the laws of

Gravity.' As a result, rocks and minerals, sediments and fossils were deposited in an orderly, lawful manner: stone and marble first and deepest, with shells in them, and the 'lighter matter, as Chalk, and the like' last.

His theory of the origin of strata had all the best attributes of a scientific hypothesis: it was testable. All too quickly, a whole range of authors familiar with the sequences of beds in coal mines pointed out that one usually found lighter layers and heavier ones sandwiched one between the other.[86] Woodward's theory also failed to explain why the Flood, with its infinite power, could dissolve granites and basalts but not delicate shells and bones, but instead preserve them perfectly and deposit them in the newly formed rocks as fossils. Nonetheless, Woodward's neat, if fanciful, scheme even has its adherents today.[87]

Woodward was not prepared to admit that the earth's surface had changed at all after the Flood – Burnet's theory standing on its head. Burnet's Flood created geological structures doomed to constant further degradation because of man's wickedness; Woodward's Flood left man with a perfect and stable world. He recognised the power of erosion: '*Rocks*, Minerals and the other elements of the Earth . . . suffer a continual Decrement and grow lower.' Nonetheless, he insisted that

> There were never any Islands, or other considerable parcels of Land amassed or heaped up, nor any Enlargement or Addition of Earth made to the Continents, by the mud that is carried down into the Sea by Rivers . . . There is no authentick Instance of any considerable tract of Land that was thrown up from the bottom of the Sea, by an Earthquake, or other subterranean Explosion, so as to become an Island, and be render'd habitable.

But the extent of erosion by water, wind and frost was too great to be ignored, so Woodward had to develop a new argument that would deal with the facts he knew as a geologist.[88] The result is an ingenious theory, parts of it quite modern, parts simply seventeenth-century sophistry within a Newtonian metaphor. Woodward's earth, post deluge, is stable but not in fact unchanging. This is possible because it is in a state of balance – a dynamic balance between opposing forces. First he envisioned a balance with respect to water: 'There is a perpetual and incessant *Circulation* of Water in the Atmosphere: it arising from the Globe in [the] form of *Vapour*, and falling down again in *Rain*, *Dew*, *Hail* and *Snow*. That the quantity of Water thus rising and falling is equal . . . yet it varies in the several Parts of the Globe.' Woodward probably borrowed this idea from Robert Hooke.[89] In fact, the subject of the earth's water was actively fought over through the second half of the seventeenth century, largely with respect to the question of what caused rain, and how rivers became recharged (where did the water come from?). Woodward, like many contemporaries, including Edmund Halley (whose name will always be associated with the famous comet), held to the view of the ancients, that rivers and springs are not recharged by rain, but from the abyss of the sea – the 'great subterranean Magazine, with its partner the Ocean, driven by internal heat'.[90]

Going one better, Woodward extended the concept of 'circulation' to the material earth itself, and on a global scale. According to Woodward this all started with the Flood and continues in the incessant deterration (wearing away) of the strata laid down at the Flood, which started a great recycling of the nutritive materials of the earth: 'The constituent Matter of any one Body being proper, and turning again thus naturally . . . there is a kind of Revolution or Circulation of it: so that the Stock or Fund can never possibly be exhausted, nor the Flux and Alteration sensible.' The Flood interred a huge amount of such organic material and erosion brings it back into use – sufficiently slowly so that the excesses of the flesh of pre-Flood days will never be repeated. The natural recycling of these 'Vegetative and Terrestrial Materials' – a sort of giant compost heap – then acts as a sort of governor, controlling the growth of nature, agriculture and population. As he put it:

> Although the *principal Intention* in the Precipitation of the Vegetative Terrestrial matter at the Deluge, and burying it in the Strata underneath the Sand, and other mineral Matter, was to *retrench* and abridge the *Luxury* and Superabundance of the productions of the Earth, which had been so ingratefully and scandalously abused by its former Inhabitants, and to cause it to deal them for the future more frugally and sparingly; yet there was a still *further Design* to that Precipitation; and the Matter so buried was to be brought up upon the Stage once more; being only reserved in store for the benefit of Posterity, and to be, by this Deterration, fetch'd out to light again to supply the Wants of the latter ages of the World.

The result, for Woodward, was a utilitarian world existing in a 'just equilibrium'.

All this circulation needed a mechanism and Woodward found it in the rain. At any one time, matter is being washed into the sea and then returned to the earth 'dispersedly by Rain'. In this astonishing leap, while he could not bring himself to allow that rain could recharge rivers and streams, he blithely proposed that it recycled all the terrestrial material eroded from the land and re-deposited it on the hills and plains (here he means both 'Vegetative' and 'Terrestrial matter'). He had painted himself into a corner, of course. Once he had decided that land could not be raised up or recreated from accumulations of sediments or by the heaving around of the earth due to earthquakes, the only mechanism left was this impossible one, evidently at odds with all empirical experience (except for phenomena like sand-storms, which do often re-deposit terrestrial material).

One other matter is worth mentioning: in line with contemporary mechanics, he needed a source of energy to drive the machine of the earth. Woodward believed that the driving force for all post-Flood changes in the earth was its 'inner heat', presumably derived as the earth cooled from its original molten state.

At almost the same time as Woodward was developing his theory in London, another comprehensive theory inspired by Burnet's ideas was being developed by a friend of Woodward, yet another Cambridge man, William Whiston (1667–1752). Whiston was Newton's chosen successor as Lucasian Professor at Cambridge, and both his disciple and a disappointment. A broad scholar, he perhaps achieved more lasting fame for his translations of the works of Josephus Flavius, the great first-century Jewish chronicler and historian, than his science.[91] He was an early patron of Samuel Clarke, a man who in 1714 was brought up before Parliament accused of lack of orthodoxy concerning the liturgy. After seven years as Professor, Whiston was himself dismissed for following the Arian heresy (of denying that Christ was God made flesh) but he could just as easily have been fired for his radical interpretations of creation in order to fit with physical theories.

It was difficult, if not impossible, to be Newton's successor. Equalling Newton's accomplishments was unlikely; the notion of surpassing them preposterous. Whatever Whiston did, it would not be enough; whatever ideas he put forward, it would be claimed that Newton had had them first, or had had better ones. Perhaps this is why he spent so much time on Jewish history. Described by his contemporaries as an intense, driven man, Whiston was far from shy and retiring and he leapt into the physico-theological arena with a sacred theory of his own,

William Whiston

far more 'modern' and 'scientific' than Burnet's or Woodward's. In 1696 he published *A New Theory of the Earth from its Original, to the Consummation of all Things.*[92]

As might be expected for a man who literally and figuratively followed Newton and was a contemporary of Edmund Halley, Whiston was enamoured of comets. He adopted comets as a causal agency to marry a freely revised biblical account of creation to a post-Copernican, post-Newtonian concept of the universe. First, like his predecessors, he had to set the stage and deal with 'chaos'. But he made a huge break with orthodox contemporary practice: not only did he consider Genesis to be a folktale, it was not even an account of the creation of the whole universe, but only of the earth: 'I prove that the History before us, extends not beyond the Earth and its Appendages, because that confus'd Mass, or rude heap of Heterogeneous

Matter, which we call the Chaos, whence all the several parts were derived, extended no farther ... Not the innumerable Systems of fix'd Stars, nor the narrow System, of the Sun, nay nor the Moon herself, but our Earth alone, was the proper Subject of the *Mosaick* Creation.'

The conservative Cambridge cleric the Reverend John Edwards quickly brought out a book of sermons against Whiston, savaging him in a fine, angry rhetoric for treating Genesis as nothing more than a fiction to bamboozle an ignorant people, and dismissing

the fond Notions which had possess'd some men's heads, and had been divulg'd by their pens, viz. that Moses is not to be understood according to the *Letter*, that he speaks to the capacity of Blockish *Brick-makers* that were recently come out of *Egypt*, and scarcely understood common sense, and therefore any Story of a Cock and a Bull would serve them ... we must not think that he would trouble the brains (if you can suppose they had any) of a company of Dull Slaves with Natural Philosophy.[93]

Whiston needed to restrict the Mosaic account to the creation of the earth alone because he was sure that he had found the source of the earth in a comet (and therefore in something with an existence prior to the events of Genesis). 'If we ... enquire what confus'd Masses or Chaos's either at present are, or ever, within the Annals of Time, were extant in the Visible World, we shall discover no footsteps of any such thing, excepting what the Atmosphere of a Comet affords us ... a Comet, or more peculiarly the Atmosphere thereof, was that very Chaos, from whence the world arose, whose Original is related in the *Mosaick* history.' This in turn meant that he had to be rather reticent about the guiding, rather than creating, role of God in any of the subsequent events.

Whiston's comet was supposed to have been diverted from

an elliptical to an inwardly spiralling orbit. It then condensed in on itself. At first, as Whiston described in a passage both surprisingly modern and equally Cartesian and Epicurean: 'The chaos was a chemical Compound of all sorts of Corpuscles, in a most uncertain confus'd and disorderly State; heavy and light, dense and rare, fluid and solid Particles were in a great measure as it were at a venture, mingled and jumbled together. The Atoms . . . every one were in every place, and all in a wild and disturbed Confusion.' Then the various parts settled out according (of course) to their specific gravities, producing a dense inner fluid, then water, with the earth's crust as the outermost surface. This new earth was essentially like an egg, but it already had 'Mountains, Plains and Vallies . . . the caverns of the Seas, with the extant Parts of the Dry Land, being in effect great Vallies and Mountains'. And, of course, high ground was essential in order for rivers to flow. Having rather shakily invoked specific gravity, Whiston was commited to using it for the cause of mountains, insisting that they must be made of lighter material than the rest of the earth. He stated that the density of the rocks on mountain tops is in the ratio of approximately 3:1 relative to water, while stones at the bottom of mines are heavier at 4:1 or 5:1, 'This proposition will also, I imagine, be new and unexpected to many.'

For Whiston, the Flood was due to a second comet that swept extremely close to the earth. He made an elaborate calculation to show that that the date of the Flood (1,656 years after Adam) coincided exactly with the path of such a comet. The atmosphere of this second comet condensed and formed vast rains, which fell for forty days and nights, and in addition it passed so close to the earth that its gravitational field caused tides of double displacement, distorting the very shape of the earth and causing the surface to fracture and release the fountains of the deep. (This was all very modern, but since the whole point

161

of the new science of comets was to show the regularity and predictability of their orbits, it negates the concept that the Flood was a more or less instantaneous response of God to man's excesses. There is more predestination due to the operation of physical laws in this formulation than Godly decision-making.[94]) A last, seemingly unnecessary, bit of 'science' thrown in by Whiston was that before this cometary near miss, the earth was in orbit around the sun but did not rotate on its axis. Only after the comet passed and the Flood broke did we acquire daily rotation and therefore the diurnal alternation of night and day. Again, this was completely at odds with Genesis and opened him up to further criticism.

The effect of the deluge in Whiston's sacred theory was to break up the upper surface of the earth and redeposit it along with all the remains of the creatures that had been living, which were then preserved as fossils. This created the present multi-layered earth: 'Our upper Earth, for a considerable depth, even as far as we commonly penetrate into it, is *Factitious*, or newly acquir'd at the Deluge; the ancient one having been cover'd over by fresh *Strata* or *Layers* of Earth at that time, and thereby spoil'd or destroy'd as to the use and advantage of mankind.' In this respect Whiston here agreed with Woodward, but he strongly disagreed with Woodward's notion that the process had involved the dissolving of the earth down to its constituent corpuscles first. That was absurd, given the vast numbers of 'Shells of Fish, Bones of Animals, Intire or Partial Vegetables, buried at the Deluge and inclosed in the Bowels of the present Earth'. After the deluge, Whiston's earth was like Woodward's but more old-fashioned – constant and unchanging: 'Since the Deluge there neither has been, nor will be, any great and general Changes in the state of the Earth, till that time when a Period is to be put to the present Course of nature.'

*　　*　　*

Paley and Darwin's rooms at Christ's College
(the two windows above the doorway).

Paley might have owned a watch like this lever watch by Joseph Emery, 1784.

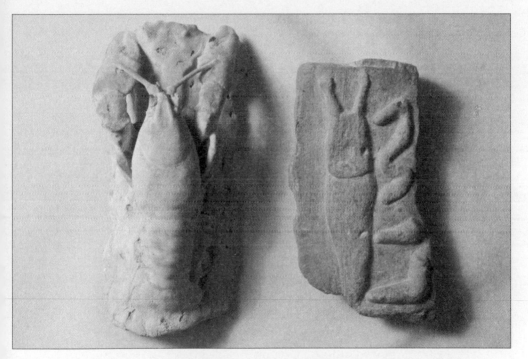

Not fossils: on the left, a modern crayfish encrusted with calcium carbonate from a natural spring; on the right, one of Beringer's 'lying stones', apparently depicting a slug.

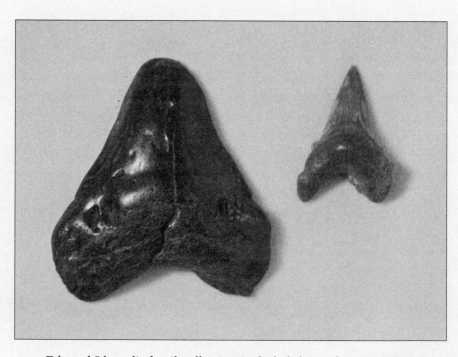

Edward Lhywd's fossil collection included these *glossopetrae* or tongue stones, now known to be shark's teeth

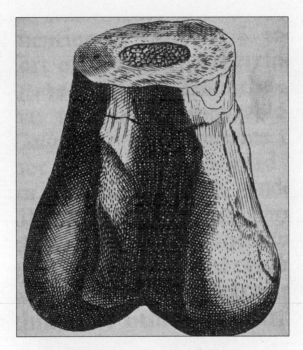

Plot's 'thigh bone of a giant'
(Natural History of Oxfordshire)
was the distal end of a dinosaur femur.

The 'Temple' of Serapis (a market), near Naples: the dark bands near the bases of the pillars are caused by the activity of marine boring organisms.

'A broken and confus'd heap of bodies': layers of folded and faulted chert in the Himalayas.

The unconformity at Jedburgh, drawn by John Clerk of Edinburgh for
Hutton's 1795 *Theory of the Earth with Proofs and Illustrations.*

For all their flaws and the outrage that they triggered within the Church, sacred theories like these gave Christian natural philosophers, among them John Ray, a figurative space within which to work. However, while it would be extremely tidy if, after 'sacred theorists' had had their say, their work triggered a new round of enquiry neatly bringing us closer to modern ideas, the opposite was true. Considerably before Burnet, Whiston and Woodward published their works, two more modern natural philosophers had pointed the sciences of the earth in a different direction: the misanthropic Robert Hooke in London and an equally unhappy Danish exile in Florence.

Hooke's *Micrographia* and *Discourse of Earthquakes* represent major milestones in the understanding of fossils and of geological processes, especially in recognising the relentless power of erosion in tearing down rocks and laying down the sequences of accumulated sediment (and fossils) that make up the earth's surface, and in accounting for the uplift of mountains by the actions of volcanoes, earthquakes and the inner heat of the earth. In a passage worthy of Ovid (the sort of language that Burnet had freely adopted), Hooke described earthquakes as 'transposing, subverting and jumbling the parts of the Earth together; overthrowing Mountains and turning them upside down'. His conclusion from all of this was that the earth was very old: '[Fossils] and the quantity and thickness of the Beds of Sand with which they are many times found mixed, do argue that there must needs be a much longer time of the Seas Residence above the same [than any biblical span].' To Hooke the empirical evidence was conclusive: earthquakes and volcanoes had played a major role in changing – often lifting up and adding to – the earth's surface, while opposing forces of erosion by wind and water wore it down and deposited the debris elsewhere. Eventually it became an article of scientific faith that these processes should not be mystical or unknowable and it

163

appeared that they were in fact the same processes that we can see acting on the earth today. Some of these processes were even measurable, for example the rate at which sediment accumulates in river deltas. In order to connect the sort of process by which the sides of a mountain valley are eroded by fast-running streams, the silt and gravel borne downstream and then deposited forming thick layers of sedimentary rocks, all the way to changes in the landscape on a gigantic scale, only *time* is needed.

In 1669, one year after Hooke's *Discourse* lecture, Steno published in Latin (with an English translation in 1671) his wonderfully named *Prodromus to a Dissertation concerning Solids naturally Contained within Solids: Laying a Foundation for the Founding of a Rational Accompt both of the Frame and the several Changes of the Masse of the Earth, as also of the various Productions in the Same.*[95] The odd-sounding title of Steno's great work needs some explanation. The story is a tortuous one, leading from the medical school in Leiden to the rocks of Tuscany. Steno began his career as a staunchly Lutheran anatomist at Leiden, where he made important discoveries (the duct of the parotid gland is still named after him). But he fell foul of academic politics and failed to get a professorship. He moved to Paris for a while but in 1665 headed for Florence, where his patron was the Grand Duke of Tuscany, Ferdinand II de' Medici, an enthusiastic supporter of medical researches and, following his father Cosimo II, the patron of Galileo. There Steno was admitted to the Grand Duke's Accademia del Cimento, and in 1666 Ferdinand asked him to prepare an account of the head of a huge shark that had been caught near Livorno. After studying this shark and its great teeth, Steno realised he was looking at something exactly like the fossils known to contemporaries as 'glossopetrae' ('tongue stones'). These could be collected all over Europe and were the same kinds of fossils that Lhwyd, Hooke, Plot and Lister had agonised

Steno

over. For Steno it was obvious that glossopetrae were not some sport of nature imitating teeth but the remains of real sharks' teeth. The first part of the *Prodromus* establishes the case for fossils brilliantly, but Steno did not stop at the debate over the nature of fossils. He had become fascinated by the deeper question of how solid objects like fossils and minerals could become enclosed in solid rock – hence the convoluted title of the short essay he dedicated to his patron.

Steno had a knack for getting straight to the processes underlying phenomena, and to do this he relied not on old-fashioned hypotheses, but the evidence of his eyes. Perhaps it was from his training as an anatomist, perhaps it was also because he had not been trained in the geological orthodoxies of the day; for whatever reason, before three years had elapsed, Steno had grasped some of the fundamentals of how the earth works. From walking the hills of Tuscany and trying to understand the structure and pattern of the rocks, Steno came to articulate what he saw as simple principles for the study of the earth. If a 'Stony Bed' is uniform in consistency it has probably been there since creation. If the stony bed is mixed in composition it was formed later. If fossils are present, then the sea must once have overflown that place. If there are coals, ash or pumice, there must once have been fire. Geological strata were all laid down horizontally at first, and then altered. And (perhaps most importantly) where beds lie on top of each other, the lower beds were formed before the upper ones and did not interfere with the formation of the latter: all devastatingly simple.

Concerned to embed his science within a deeper philosophy and the physics of matter and motion, Steno asserted the universal acceptability of his new theories:

> For what I have affirmed of *Matter* hath place every where, whether you take Atoms for your Matter, or Particles a thousand way variable, or the Four Elements, or Chymical Principles never so much varied according to the variety of Chymists. So also what I have proposed of the *Determination* of Motion, agreeth with every Mover, whether you make it to be the Form, or the Qualities flowing from the Form, or the Idea, or the common subtile matter, or the proper subtile Matter, or the particular Soul of the World, or the Immediate Concourse of God.

While it is largely independent of any scriptural 'facts', his account of earth history is in its way the first of the sacred theories. Indeed, he was very careful to reassure his readers, not least of them Grand Duke Ferdinand, that: 'Least [sic] there should be apprehended any danger in the novelty, I shall in short lay down the agreement of *Nature* and *Scripture*.' And it was accepted as such; the last page of his work is a certificate pronouncing it to be acceptable in the eyes of the Church.

Steno's *Prodromus* is the first synthesis of empirical investigations, in a real geological setting, in terms of law-like cause and effect, and is perhaps the more remarkable when one remembers that it was written only thirty-five years after Galileo's appearance before the Roman Inquisition. He used his 'principles' to read – as if from an open book – the history of rocks around Tuscany. No one had done anything like that before. Starting with the simple strategy of distinguishing two main kinds of rocks, 'stony' and 'sandy (including clays and small stones)', Steno identified six stages in the history of the earth. In the first stage, the earth was covered by sea. There was no life and there were no 'heterogeneous bodies' of rock (no strata). This obviously corresponds with verses one through eight of Genesis 1. In the next stage, the sea had receded underground and now the earth existed as 'plane and dry' with the original stony rocks exposed at the surface (Genesis 1:9). However, as a result, vast cavities formed underground without causing the surface to break. Next, due to the collapse of these subterranean caverns, the stony surface was broken up, creating mountains and valleys (these are the mountains and valleys mentioned in the Bible pre-Flood; the process is from Descartes).

Now the earth was covered with water again. The Biblical Flood submerged the earth, during which event new (sedimentary) beds of sands and clays were laid down covering up these valleys. Following this, the upper, dry surface began to erode as the seas receded. The underlying beds were consumed, again without the surface breaking. Finally, in the sixth stage, the superficial sedimentary rocks collapsed inwards once more, forming a new landscape of mountains and valleys in which both the new ('sandy') and older ('stony') rocks were exposed. Here Steno admits there was no biblical authority and thereafter he concluded that the earth continues to change through erosion and earthquakes. 'Hence it is, that we may distinguish Six distinct casts of the Country of *Etruria*, and that it hath been twice Fluid, twice Plane and Dry, and twice scabrous and Craggy' – a single history which 'I confirm . . . to be true of the whole Earth'.

Steno saw that erosion was a major factor in shaping the earth '[It] is certain, that a great parcel of Earth is every year carried into the Sea [and] does leave new lands fit for new inhabitants.' But the driving force in his geology is inner heat and the creation of great underground caverns, whose collapse throws the otherwise horizontal beds into disarray: 'The force of Subterranean Fires [was] made more rapacious . . . [giving] place to greater ruines . . . violent excursions of the Beds . . . spontaneous falling down of the upper-beds when the lower matter or foundation [was] withdrawn . . . [The] bottom of the sea [was] rais'd up by the dilated caverns under the earth . . . upwards, whether that be caused by a sudden accession of underground Exhalations, or by a forcible elision of Air . . .'

If these fiery subterranean caverns smack more of Dante's *Inferno* than sober geology, we should not lose sight of the modernity of Steno's ideas. Burnet, Whiston, Ray, Plot, Lhwyd, Lister and all the others came after Steno so it is more than ironic (but not atypical of the way ideas progress) that while Steno had got things largely right, the other sacred theorists, bound up in their own religious premises, committed to their own solutions and ignoring the more incompatible elements of contemporary science, set off in different directions.

As for Steno himself, his later career involved yet more abrupt turns. The *Prodromus* was supposed to be a mere outline of a great geological work, but that was never finished and there is no evidence that it was really started. Having grown up a Lutheran, he seems to have been repelled by the authoritarianism of Danish Protestants, but then in 1667 converted to Catholicism and took holy orders, apparently comfortable in the shadow of the Inquisition. His early career as an anatomist was founded not just on his skill but also the virtuosity of his public dissections. His geology was more secretive and introspective, perhaps because he saw how close he had come to challenging the biblical account of creation. When a professorship back in Holland was finally offered his Catholicism obviously became an issue. He returned home but soon gave up his academic career completely to become a bishop, ending his career ministering to the outcast Catholic minorities of northern Europe, where he died, worn out and unsung, at the age of forty-eight.[96]

While Steno was roaming the hills of Tuscany, putting together the evidence for his theories, John Ray was just beginning to assemble his own views of natural theology and to preach them in sermons. Thirty years later, having spent the bulk of his career as a cataloguer and organiser of nature, and with his classic *The Wisdom of God* complete, Ray turned his attention to the geology and the physico-theological theorists. None of them – not Woodward, not Whiston, and most definitely not Burnet – found favour with the 'father of English natural history'. But Hooke he could not ignore. In 1693 Ray produced the first of many editions of a book dealing more with the physical world: *Three Physico-Theological Discourses*. Like others before and since, being only too aware of the threats that natural philosophers posed for orthodox Christianity, Ray spelt out the arguments of these enemies from without, and answered them. However, for all that his research in natural science was forward-looking and innovative, his theological world view was far more conservative. In reading Ray we might wonder whether, in the long view, it was a step backwards rather than forwards, a defence rather than the capture of new ground.

For Ray there could be no flirting with Descartes' vortices and other atomistick heresies. For him, the history of the earth was written truly in Genesis. And where Burnet thought that 'the present Earth looks like a Heap of Rubbish and Ruins', Ray protested that 'the present Face of the Earth, with its Mountains and Hills, its Promontories and Rocks, as rude and deformed as they appear, seems to me a very beautiful and pleasant Object, and withal that variety of Hills, and Valleys, and Inequalities, far more grateful to behold, than a perfectly level country, without any Rising or Protuberancy, to terminate the Sight.' This is utilitarianism on a global scale:

A Land distinguished into Mountains, Valleys, and Plains [is] useful to Mankind in affording them convenient Places for Habitation, and Situations for Houses and Villages, serving as screens to keep off the cold and nipping Blasts of the Northerly and Easterly Winds, and reflecting the benign and cherishing Sun-Beams . . . promoting the Growth of Herbs and Trees . . . also most convenient for the entertainment of the various Sorts of Animals, which God hath created, some of which delight in cold, some in hot . . . places.

Ray also took up the issue of the causes of rain: among them 'Use and Necessity of the Mountains and Hills [was] the Maintenance of Rivers and Fountains . . . and for the Generation of Rivers and Fountains which in our hypothesis that all proceed from Rain-water could not be without them'. It is a mixture once again of the old and the new, of innovation and compromise.

Ray's version of a sacred theory to explain the structure of the earth is a curious mixture of First and Second Causes, supported equally by quotations from the Bible and from ancient and modern writers. His first stage is orthodox with a twist: out of chaos, God created 'the Earth or Terraqueous Globe' in an orderly form but with each creature containing the seeds of all subsequent living creatures. Next was 'the Separation of the Water from the Dry Land and Raising up of the Mountains'. After this, God's seeds were able to hatch out (as in Genesis 1:20–25, 'Let the Waters bring forth abundantly the moving Creatures that hath Life, and Fowl that may fly above the Earth.') After the peopling of the earth, 'God [gave] to every Species a power to generate or propagate its Like.' But, if the first form of the earth was watery, what caused the dry land and what made the mountains rise up? Here Ray agreed with Hooke that it was due to earthquakes and 'Subterranean fires'. Supporting this view, he massed dozens of historical accounts. For example, 'the greatest and highest Ridge of Mountains in the World, the Andes of Peru, have been, for some hundreds of Leagues in Length, violently shaken, and many Alterations made therein by an Earthquake in the Year 1646.' And furthermore, earthquakes and volcanoes should be seen as good because they allow the earth to vent the pressure of fire and gas building up internally.

With respect to the Flood, Ray sided with Hooke: it was but one of many such inundations, none of them global in extent and none the cause of major physiographical features or geological structures of the earth (and most definitely not in the way either Woodward or Burnet hypothesised). To account for Noah's Flood, he proposed a shift in the centre of gravity of the earth: 'Whereupon the Atlantick and Pacifick Oceans must needs press upon the subterranean Abyss, and so by Mediation thereof, force the Water upward . . . breaking up the Fountains of the great Deep.' But he does not say what caused the shift, perhaps for fear of seeming to side with Whiston; instead he supposes that 'some Divine Power might at that time, by the Instrumentality of some natural Agent, to us at present unknown, so depress the Surface of the Ocean, as to force the Waters of the Abyss.'

In *The Wisdom of God*, Ray had tended to the view that the earth had been relatively stable 'since the ancient Times recorded in History'. From this it had to follow either that 'the World is a great deal older than is imagined or believed' or that the very earliest history of the earth had been particularly unstable, suffering 'far more concussions and mutations' than more recently. In *Three Physico-Theological Discourses* he allowed that the earth 'is in a violent State' that will eventually subside as the world comes to an end. Departing from his trademark utilitarianism, Ray concluded that the earth was not stable but rather, like the despised Burnet, he saw forces of change everywhere. He concurred with Hooke and his friend Edward Lhwyd in recognising the power of the elements in wearing down mountains and turning them to mud: 'the Rains continually washing down and carrying away Earth from the Mountains, it is necessary that as well the Height as the Bulk of them . . . should answerably decrease'.

While Hooke and others proposed that sedimentary deposits were eventually uplifted to form new land, Ray insisted that any new land created by the deposition of eroded silt remained at sea level. Reluctantly, therefore, he was forced to acknowledge the likelihood that these natural processes would eventually level all the mountains and compel 'the Waters to return upon the dry Land, and cover the whole Surface of it, as at first . . . Land and Water [returning] to their ancient and primigenial State and Figure'. But while that had scriptural authority ('Every valley shall be exalted, and every mountain and hill shall be made low', Isaiah 40:4), it was too close to Burnet's vision, so Ray tried to leave open the possibility that it was wrong: 'I cannot imagine or think upon any natural Means to prevent and put a stop to this Effect, yet I do not deny that there be some.'

Ray could not allow the final connection: that if mountains are both built up (by earthquakes and inner fires) and worn down (by rain and frost), these tendencies could be in balance. But he was no doubt forced by his theology to resist a balanced cycle that would create a changing but eternal earth – the Bible foretells an End. Almost a hundred years would go by before James Hutton would follow Hooke in making just that leap of understanding.

Ray died in 1705, whereupon his friend and protégé William Derham continued his work, preaching Ray's version of physico and natural theology and continuing to pour scorn on 'our eloquent theorist' Burnet. The explanatory value of physico-theology soon declined but, eschewing the worst excesses of Ray's utilitarianism, Derham brought in more data for natural theology which, for the next hundred years or so, continued to present the ideal means of marrying science and theology, thus ensuring that everyone could remain on the same side – God's. This symbiosis lasted well into the middle of the eighteenth century until it in turn began to crumble under the weight of new scientific discoveries and even more radical theories.

CHAPTER EIGHT

Unfinished Business:
Mountains and the Flood

'The Truth is, the Source of Nature has been already too long made only a Work of the *Brain* and the *Fancy*: it is now high time that it should return to the plainness and soundness of *observations* on *material* and *obvious* things.'
Robert Hooke, *Micrographia*, 1665

On 28 February 1831, HMS *Beagle*, with Charles Darwin on board as naturalist, reached the South American coast. Over the next three years, Darwin found evidence everywhere that the eastern coastline of the continent had been raised some 20 to 100 feet. Vast stretches of the coast consisted of plains of pebbles that clearly had been under the sea not many thousands of years before, as evidenced by the fact that they were peppered with the remains of modern oyster species. After three years of relentless pressure, in 1834 the hard-driving FitzRoy allowed himself a small treat: an expedition to the headwaters of the river Santa Cruz. A party of twenty-five men took three boats and made their way upstream until they were only sixty miles from the Pacific Ocean and could see the Andes far off to the west. The first part of the journey was across this great plain of stones. Darwin saw it as an ancient seabed and as conclusive evidence of a recent uplifting of this section of the coast. FitzRoy agreed, as he later regretfully wrote: 'While led away by sceptical ideas, and knowing extremely little of the Bible, one of my remarks to a friend, on crossing vast plains composed of rolled stones . . . some hundred feet in depth, was "this could never have been effected by a forty days' flood".'[97]

By early 1835, the *Beagle* had worked her way round to the west coast of South America and, while her officers were surveying the region around Concepcion in Chile, Darwin set off up into the Andes, among other things collecting for himself examples of that great enigma, fossil shells preserved high above the ocean level. (He also came down with a mysterious, near-fatal fever.) On 20 February 1835, as the *Beagle*'s officers continued their laborious but exquisitely accurate mapping of the coastline and harbours the region around Concepcion experienced a major earthquake and tidal wave. Captain FitzRoy wrote in his log: 'Suddenly an awful overpowering shock caused universal destruction – and in less than six seconds the city was in ruins. The stunning noise of falling houses; the horrible cracking of the earth; the desperate heart-rending outcries of the people . . . can neither be described nor imagined.'[98] From the ensuing tidal wave 'the body of water reached twenty-five feet above the usual level of high water'. As the *Beagle*'s officers had just surveyed Concepcion harbour, there was no doubting the evidence of their eyes: 'the southern extreme of the island was raised eight feet, the middle nine, and the northern end upwards of ten feet'. This seemed to Darwin to be a perfect example of Hooke's views, extended by Lyell, on the role of earthquakes in lifting up the earth.

Two issues important to the history of natural theology have been left unresolved and we should give these at least a moment's attention before returning to natural theology and evolution because they are emblematic of the ways in which natural science and theology began to diverge. They concern mountains and the Flood. This means we must look at the work of Dr James Hutton, the Scottish polymath physician and farmer who gave geology and the history of the earth its current, distinctly non-biblical, structure. Up to now, this story has mostly been told in terms of the English element of the Age of Enlightenment. James Hutton, along with David Hume and Adam Smith, was one of the fathers of the Scottish Enlightenment.

Hutton (1726–97) was the son of William Hutton, a wealthy merchant and one-time City Treasurer of Edinburgh who died when James was very young. In his early years as a student at Edinburgh, Paris and Leiden, James delved into chemistry before turning to a more conventional subject, the law. Changing again, he studied medicine and eventually graduated as a doctor, although he never practised. Like Hume and Darwin, he was a philosopher at heart and his intellectual interests were broader than conventional professional training allowed. Returning to Edinburgh he took up a scientific study of farming, learning methods directly from travels and studies in Norfolk, Holland and France. During this time he became a fairly serious student of mineralogy. In 1754 he settled back on the family estate, where he put his ideas into practice and also ventured into businesses in coal and the manufacture of sal ammoniac that established him financially. There are some hints of his contemporary Erasmus Darwin here, except that Hutton was a lean, spare, reserved man, sober in his dress and reserved in his manner – almost a caricature of the dour Scot. He never married but lived most of his life with his three spinster sisters. He was a deeply learned man with a driving interest in everything we would now call natural philosophy, and took a leading role in the Edinburgh Philosophical Society, which later became the Royal Society of Edinburgh. But nothing in the early career of this intensely practical man gives a hint of the intellectual bril-

James Hutton

liance lying behind his first venture into geological theory. Perhaps, like Steno, Hutton was sufficiently isolated from the geological orthodoxies, fashions and squabbles of the day to be freer to speculate than his contemporaries among the English and Scottish academics. He was certainly well-enough read in the relevant literature and had a wealth of practical knowledge on which to found his own opinions. Like Steno he had walked the land (in 1747, for example, he made a walking tour of England with the inventor James Watt) and had observed intently. Through the mind of Hutton, the results of the previous 150 years of detailed analysis of the earth – especially where scholars had not been encumbered by authorities such as Aristotle and the Bible – were synthesised into an essentially modern, secular theory of the earth.

It began with the sort of chaos that Descartes and Whiston were comfortable with, involving an originally molten earth cooling down and only later acquiring the oceans. This origin occurred many years ago (we now think 4.5 billion years ago). The geological record of the last two billion or so years tells of the slow emergence of ever more complicated forms of life, with humans the most recent and perhaps most vulnerable of these, having been around only for a few hundreds of thousands of years (depending on definitions of 'human'). As the work of people like Steno, Hooke and a host of others had multiplied, some things had become patent early, some held their secrets much longer. As we have seen, quite an early observation was the fact that the action of the weather – water, frost, wind – serves to break down what otherwise appears to be hard and unyielding rock, slowly eroding it to gravel and sand and eventually building up new sediments in the water or soil on land. As the great eighteenth-century French natural philosopher Georges Louis leClerc, Compte du Buffon, put it, the processes that eroded mountains into sand and built up new land, 'operations uniformly repeated, motions which succeed one another without interruption, are the causes which alone ought to be the foundation of our reasoning'. Hutton extended this 'uniformitarian' principle to say that the processes acting on the earth now are the same as those that have shaped it in the past. For him the history of the earth was not explained by unique, supernatural, cataclysmic events, but by the steady drip of everyday phenomena. And from this, although it was even less easy to assimilate, he (like so many others) read the consequence that, because these processes act extremely slowly, the earth must be very old.

The key to the new theory was the concept of 'cycling'. As early as 1668, Hooke realised that there must be some process by which strata were uplifted, only to be torn down once more.

Remarkably, he used a cosmological analogy: 'In the circular Motion of all the Planets, there is a direct Motion which makes them endeavour to recede from the Sun or Center, and a magnetick or attractive Power that keeps them from receding. Generation creates and death destroys ... All things circulate.' This is 'remarkable' because history has usually credited Newton with this theory of orbital mechanics, but Newton's publications on that subject were in the distant future. In fact, Hooke (who often seemed to become embroiled in arguments over priority for ideas) has a good claim to have been the first to develop this explanation of planetary orbits, having presented the idea in a 1666 lecture to the Royal Society. Only eighteen years later did Newton provide the mathematical basis for this principle demonstrating in the *Principia* the key fact that the inverse square decrease of the gravitational force with distance accounted for the elliptical shape of planetary orbits that Kepler had already demonstrated.

Hooke's *Discourse of Earthquakes* (1668) contained one of the very first intimations of a cyclic theory applied to earth processes.[99] Hooke realised that there was a balance of forces: while the geological strata were being formed and mountains were raised up, at the same time the land was constantly being eroded, 'washing down the tops of Hills, and filling the bottoms of Pits'. As usual, the problem was to discover how this cycle might be driven.[100] For Hooke, the principal cause of mountain uplift was earthquakes, which also lifted up land from the bottom of the sea and caused other parts of the dry land to sink. Other causes were the heat at the centre of the earth and possibly a shift in the earth's axis of rotation.

Similarly, in 1704, the atheist-pantheist writer John Toland expanded on a vision of all matter recycling over time, deriving this view from a catholic combination of Epicurean atomism, Cartesian vortices and Newtonian motions. According to him, the whole universe was a series of systems that physically recycled within and among each other:

> Tho the Matter of the Universe be every where the same, yet, according to its various Modifications, it is conceiv'd to be divided into numberless particular Systems, Vortices, or Whirlpools of Matter, and these again are subdivided into other systems . . . the earth . . . is divided into the Atmosphere, Ground, Water, and other principal parts; these again into Men, Birds, Beasts, Trees, Plants, Fishes, Worms, Insects, Stones, Metals . . . these depend in a Link on one another, so their matter . . . is mutually resolved into each other . . . in a perpetual Revolution; Earth becoming Water, Water air, Air Aether, and so back again in Mixtures without End of Number. The Animals we destroy contribute to preserve us, till we are destroyed to preserve other things, and become grass . . . All the parts of the Universe are in this constant Motion of destroying and begetting . . . [As previously noted, all this still required and necessitated the guiding hand of God.] . . . All the Phaenomena of nature must be explain'd by Motion, and the action of all things on one another, according to mechanick Principles, and 'tis so in effect that they account for all the Diversitys in Nature.[101]

In 1785, Hutton presented a paper to the new Royal Society of Edinburgh outlining his theory of an earth system and synthesising ideas of cycling in the physical earth in a more or less modern form (the paper was actually read for him by his friend the influential chemist Dr Joseph Black). It has been described as the 'last great philosophical masterpiece to come out of Edinburgh'.[102] In Hutton's theory (expanded in 1788) the earth, ever since it attained anything like its present state, has constantly been in a dynamic equilibrium of change of the general sort that Hooke had envisaged. Natural forces erode the land and carry it slowly to the sea. The land is also constantly, if imperceptibly, raised up. Volcanoes sporadically spread lava and ash over its surface. Sediments, having been buried, are slowly compressed, baked and converted into hard rock, and with them any remains of organisms are equally slowly turned into fossils by slow but inexorable processes of chemical change. The complex patterning of the various strata that make up the earth's surface is explained by the relentless application of these forces. Contrary to the linear theories that aimed eventually at some God-driven Final Conflagration, for Hutton the evidence of the rocks demonstrated a cyclic history powered by Newtonian steady-state dynamics: the more it changed, the more it stayed the same. He argued that: 'We are . . . led to see a circulation in the matter of the globe, and a system of beautiful oeconomy in the works of nature. This earth, like a body of an animal, is wasted at the same time as it is repaired. It has a state of growth and augmentation; it has another state, which is that of diminution and decay.'[103] (It is worth noting that Hutton's medical dissertation was on the circulation of blood.[104])

Hutton's earth is in a constant state of flux due to processes acting over millions of years as mountains are eroded by rain and frost. In turn, the steady raising up of mountains, balances their steady reduction through erosion. So far, so conventional. But Hutton's field observations showed him successive creations and re-creations of the world. Just as one has replaced the other, our present world will eventually be destroyed in its turn. This is not Burnet's single narrative punctuated by destruction of the Mosaic world at the Flood and ending at the Final Conflagration, but a steady process that is neither self-limiting nor limited from elsewhere. Hutton saw the earth very much as a machine, likening the creation of new land to replace the old as 'a reproductive operation, by which a ruined constitution may be again repaired and a duration or stability thus procured to the machine, considered as a world sustaining plants and animals'.[105]

Hutton neatly glossed over the problem of Genesis by allowing the recent creation of man, but an ancient history for the rest of nature:

> The Mosaick history places the beginning of man at no great distance; and there has not been found in natural history, any document by which a high antiquity might be attributed to the human race. But this is not the case with regard to the inferior species of animals ... we find in natural history monuments which prove that those animals had long existed; and we thus procure a measure for the computation of a period of time extremely remote, though far from having been precisely ascertained.[106]

As a deist, Hutton found all change purposeful and his theory has strong echoes of Ray and Woodward. It is a vision of a world constantly being replenished for man's uses. He had no problem in assigning almost everything to Second Causes without abandoning teleology. The orderliness of geological change was more evidence of God's design. In the first version of his theory he put things simply and elegantly: 'According to the theory, a soil, adapted to the growth of plants, is necessarily prepared, and carefully preserved; and, in the necessary waste of land which is inhabited, the foundation is laid for future continents, in order to support the system of this living world.' In a second, longer version (published in 1788), this passage became: 'The globe of this earth [is considered as] a machine, constructed upon chemical as well as mechanical principles, by which its different parts are adapted, in form, in quality, and in quantity, to a certain end; an end attained with certainty and success; an end from which we may perceive wisdom, in contemplating the means employed.' Earth was made for man, therefore; the whole process of creation and destruction of the earth in its apparently unending cycles is directed at maintaining a soil suitable for human life.[107]

The first version of Hutton's theory seems in retrospect to have been an essentially theoretical exercise. It did not cause a stir and he realised that he would need both to develop the theory and to flesh it out with a mass of supporting empirical evidence. So he set out on a series of excursions across Scotland, variously in the company of Dr Black and the mathematician Dr John Playfair, who later became the champion for his theories. There he found all the evidence he needed of an earth that had been drastically restructured over the ages. In particular, at Jedburgh and again at Siccar Point he saw perfect examples of one (upper) layer of horizontal sandstone lying unconformably on top of a much older (lower) stratum of schist that had previously been thrown into a vertical orientation, the intervening sequences having been eroded away before the upper layer was deposited. This could not possibly be interpreted as evidence that the earth's crust had been laid down at a single event, or in one continuous process, let alone over a short period of time.

By 1795, Hutton's tiny, beautifully written *Abstract* had been expanded into a two-volume treatise, *Theory of the Earth with Proofs and Illustrations* – more assertively titled, more instructive, but alas less stylish. Hostile critics even said it was too difficult to follow. (The calumny that Hutton could not write is perpetuated by many modern commentators who evidently have not read the *Abstract*.) Even in this final version of his theory, which is essentially a textbook of geology, Hutton continues with the deist, utilitarian theme: 'Why destroy one continent in order to erect another? The answer is plain; nature has contrived the productions of vegetable bodies, and the sustenance of animal life, to depend . . . [on] destruction of one continent . . . renovation of the earth in the production of another . . . Thus a world peculiarly adapted for the purpose of man, who inhabits all its climates, who measures its extent, and determines its productions at his pleasure.'[108]

Hutton's greatest contribution was to see more clearly than anyone else, and to argue more convincingly, that immense time was needed for the processes that form and reform the earth. If he was more successful than others in preaching this message, it must in part be due to the consistency of his argument, but also to the accumulation of a hundred years of evidence and debate. Like Charles Darwin later, Hutton produced an idea whose time had come. But Hutton was also cautious about specifying the exact length of time the earth had been evolving.

Hutton's original motive in studying geology had been as a classic exercise in natural philosophy, an attempt to determine an age for the earth scientifically by measuring the rates of geological processes. He concluded, however, that one could not measure the rate at which land had been lifted up, therefore the only chance of measuring the age of the earth would be in extrapolation from the rate at which it is, observably, being wasted away. But this was also impossible, given the state of contemporary knowledge in 1795:

The natural operations of this globe . . . are so slow as to be altogether imperceptible to men who are employed in pursuing the various occupations of life and literature. We must not ask the industrious inhabitant for . . . the origin of this earth; he sees the present, and he looks no farther into the works of time than his experience can supply his reason . . . *it is in vain to seek for any computation of the time, during which the materials of this earth have been prepared in a preceding world, and collected at the bottom of a former sea*' (emphasis added).[109]

In the preliminary *Abstract* of 1785, Hutton had concluded the issue rather more dramatically: 'As there is not in human observation proper means for measuring the waste of land upon the globe, it is hence inferred, that we cannot estimate the duration of what we see at present, nor calculate the period at which it had begun; *so that, with respect to human observation, this world has neither a beginning nor an end*' (emphasis added). In the larger 1788 version, the language had changed and he had added a new layer of conclusion, the last sentence of which contains perhaps the single most elegant phrase of any British scientist, of any age:

> Here are three distinct successive periods of existence, and each of these is, in our measurement of time, a thing of indefinite duration . . . in nature there is wisdom, system, and consistency. For having, in the natural history of this earth, seen a succession of worlds, we may from this conclude that there is a system in nature; in like manner, from seeing revolutions of the planets, it is concluded, that there is a system by which they are intended to continue those revolutions. But if the successions of worlds is established in the system of nature, it is in vain to look for anything higher in the origin of the earth. The result, therefore, of our present enquiry is, that we find *no vestige of a beginning, – no prospect of an end*.[110]

'No vestige of a beginning, no prospect of an end'; Hutton did not mean that there *was* no beginning or end, just that any evidence for them was irretrievably buried in the rocks.[111] But how sad it is that when he wrote everything out in the form of a textbook, his lovely last sentence had disappeared!

In 1794, while working on his two-volume text, Hutton also published a huge treatise on epistemology and metaphysics but that too fell on stony ground. He did not live long enough to

Charles Lyell

see his geology vindicated; he died in 1797. In 1802 his colleague John Playfair brought out a book that explained Hutton's ideas and finally brought them the readership and recognition they deserved. Thirty years later, Charles Lyell gave us essentially the outlines of modern theory in the book that Darwin took with him on the voyage of HMS *Beagle*, and with its help became first a geologist and then an evolutionist.

Hutton and Lyell together envisaged a uniformitarian world, but the earth has not always proceeded along its path in so imperceptibly stately a fashion as they imagined. Certain periods of its history have been more violent than others. Some processes were slower, formerly, and some no doubt were faster. For example, in recent history a lava flow has meant something quite local in scale, as on Maui in the Hawaiian Islands. But at the end of the Cretaceous, some 65 million years ago, molten rock flowing from the earth's mantle burst upwards and covered

186

an area equivalent to most of modern India. Its remains in western India (the Deccan traps) still extend over some 200,000 square miles, in places up to a mile and a half thick. This is not evidence of a new divinely driven catastrophism of the sort that Steno had relished, however, just a reminder that we have always to admit with Hamlet that 'there are more things in heaven and earth . . . than are dreamt of in your philosophy'.

In all that has been written in the preceding chapters about geology, there is one glaring gap. The great age of the earth had become indisputable but still no one had fully explained how huge portions of the earth could become lifted up from the bottom of the sea, or how they became broken asunder, or how mountains were formed, or how volcanism could occur on the scale of the Deccan traps. Interestingly enough, Hutton, like Steno, had not accepted the earthquake explanation of mountain formation (orogeny), reverting to older theories about the expansive power of the earth's inner heat. But he knew of no method through which this inner heat could be maintained or how it acted to move the earth around. Another problem was that mountain-building seemed to be cyclic; so was the inner heat at times turned off, then on again? That seemed even more improbable.

There was a whole other theory, one in which the force that shaped the early earth was not the heat of a cooling mass of fiery sun-like material, but water. The fiery theory was termed 'Vulcanism' and the watery theory, naturally enough, was called 'Neptunism'. A Neptunist theory was much more in keeping with the account in Genesis of an initial watery earth consisting of 'the deep' and later being shaped by the great Flood, and could therefore act as a bulwark against the Cartesian atomists.

The high priest of Neptunism was Abraham Gottlieb Werner (1749–1817) of Freiberg University. Werner was a mineralogist who used his knowledge of the composition of rocks to make many brilliant contributions to the analysis of earth structure. But he had the fixed idea that all the forces shaping the earth were water-driven. In Werner's Neptunism, mountains had never been raised up at all, but rather the seas had diminished and, over the years, erosion had cut down the dry land differentially. In this view, which has certain parallels with Leonardo da Vinci's and Steno's, mountains were simply the remnants of an old flat surface of the earth.

The plausibility of Neptunism, and its obvious references to the biblical Flood, were part of the orthodoxy against which Hutton had contested. However, by the 1820s it was clear to most geologists (except most notably Robert Jameson, who taught Charles Darwin geology at Edinburgh between 1825 and 1827) that one could not explain the origin of all minerals or all rocks (basalts, for example) by watery processes. Water was important in shaping the earth, but not the only or even the dominant factor. Vulcanism prevailed as the great engine driving the changing earth, but Hutton's explanation of the cause was swept aside. While most geologists came around to the view that mountains truly had been raised up, with Charles Lyell they followed Robert Hooke and his *Discourse of Earthquakes*. Example after example seemed to confirm empirically the power of earthquakes. In 1831, Charles Darwin seemed to have had first-hand proof of it at Concepcion. Nevertheless, no one had managed to answer John Ray's perfectly reasonable objection – that earthquakes do not always occur where there are mountains. Only in our own lifetime – in fact, since the late 1950s and early 1960s – has a new and dramatically more powerful theory of the earth begun to emerge. It is a theory with great explanatory power, and is worth setting out here to see how close some of these early geologists got.

The essence of the modern theory can be seen in any aerial

view of the Appalachian Mountains or the Himalayas. They clearly were not pushed up by a local earthquake, instead the whole earth has been thrown laterally into folds as if one had pushed the skin on a rice pudding. Some vast force acting more horizontally than vertically has pushed up our mountains. Ironically, Burnet may have been closest of all to the cause of mountains in one limited sense – the language he used. The wholesale upheaval and destruction of the earth's surface, witnessed in its shattered, broken, baked, frozen, folded, spindled and mutilated state, had to have been caused by something operating on a grand global scale. The true cause of mountain-building, and also of earthquakes and volcanoes, is a phenomenon known as plate tectonics. At its simplest it derives from the fact that under the superficial crust the earth's mantle is semi-liquid. It is hot and fluid enough to move (through convection cells), carrying with it chunks of the crust, which are called 'plates'. These plates ride around, smashing into each other and riding up one over the other at the geologically breakneck speed of several centimetres a year. As a result of the churning of the mantle, in places where plates collide the crust is either thrust upwards (building mountains) or sucked down into the molten parts. Where plates are separating, new magma is forced up to the surface. The result is that the plates move around, become buckled, fractured, subducted again. Earthquakes and volcanoes reshape the land, huge faults appear – all fundamentally driven by 'inner heat'.

That wonderful polymath Benjamin Franklin was one of those who first thought of this mechanism: 'I imagined . . . that the internal part might be a fluid . . . [and that the solid crust] might swim in or upon that fluid. Thus the surface of the earth would be a shell, capable of being broken and disordered by any evident movements of the fluid on which it rested.'[112] Alfred Wegener, at the beginning of the last century, made the crucial observation that the coastlines of the major land masses can be fitted together like a giant jigsaw; therefore, they must once have been joined up. Although people like Wegener and their theories of 'continental drift' were shunned and ridiculed, the notion that, say, Africa and South America (which fit like a hand in a glove) were once physically connected is also supported by many similarities in their flora and fauna. In fact, the Americas are steadily drifting away, year by year, millimetre by millimetre, from Europe and Africa as new material is continuously forced up through the mid-Atlantic Ridge (mostly submarine except in a chain of mid-oceanic islands running from Iceland to the Azores, to Saints Peter and Paul (São Pedro and São Paulo), to Ascension and Tristan da Cunha). India is being forced relentlessly into the Asian plate, causing the Himalayas to rise up. The North American plate is moving laterally with respect to the Pacific plate, causing earthquakes along a series of faults, the most famous of which is the San Andreas system. In the 1930s, Arthur Holmes worked out the geology of the mechanism and the fact that the earth's inner heat (like that of the sun) comes from atomic fission. The clinching evidence came in the 1950s and Wegener was vindicated. This is almost always the way in science (and most of the rest of human affairs); it certainly was the case with Darwinism. For mountain-building, as for evolutionary change, evidence had been around for 200 years or more, but in the absence of a realistic and testable mechanism, other ideas – more comfortable for the status quo and for orthodoxy – had prevailed.

The second piece of unfinished geological business concerns Noah's Flood. Scholars throughout the last 500 years or more have been fascinated by the subject and have sought material evidence confirming the biblical account and thus vindicating a belief in the literal truth of Genesis. Endless expeditions have been sent in search of relics of the ark that Noah supposedly built.[113]

Although careful reading of Genesis shows that the biblical Flood, if it actually happened, was confined to a small area of the Middle East, many eighteenth- and early-nineteenth-century geologists, taking the biblical account seriously, had no difficulty in finding a wealth of evidence that they thought was compatible with the proposition of the great Flood extending across Europe. Western Europe in particular is literally covered in the geological consequences of water-driven processes on a huge scale, acting (in geological terms at least) at considerable speed. Further afield, as already noted, Robert FitzRoy interpreted the plains of 'water-worn shingle stones' that he and Darwin had seen on their Rio Santa Cruz exploration in 1834, as 'diluvial accumulations' and evidence for the Flood.[114] Everywhere in both the northern and southern hemispheres there is evidence of valleys recently carved out by water. In many places one can find boulders ('erratics') that have clearly been moved from a source many tens, or even hundreds, of miles away. The layers of sand and gravel that seem to mark these major aqueous events also mark a quite sharp transition in fossil faunas. Before (below) the 'diluvial' sands and gravels there are more tropical fossils; after (above) them we find a more or less modern fauna. No more lions in London, no more woolly rhinoceros in Oxford or hyaenas in Yorkshire, no more sabre-tooths in California.

Hyaenas turn out to be central to this particular case and with it we revisit the Reverend Professor William Buckland, variously Reader and Professor at Oxford, Canon of Christ Church and Dean of Westminster. Among other distinctions, Buckland was the world's first experimental palaeontologist, a distinguished cleric, a brilliant teacher, and a noted eccentric. No other Oxford don, as far as I can discover, has kept a hyaena in the garden. Few, if any, have been so assiduous in plying

191

William Buckland

their guests with dishes devised as a gastronomic tour through the animal kingdom. Dining *chez* Buckland meant sampling beaver, crocodile or badger (though he never managed to find decent recipes for bluebottle flies or moles). In all this, students, colleagues and family seem to have been remarkably tolerant, even supporting and encouraging. Mary Morland Buckland was an accomplished artist who accompanied her husband on his field trips; many of her beautiful watercolours of geological scenes survive, among them one that a devoted young student, John Ruskin, turned into a finished watercolour (now in the collection of the Ashmolean Museum).

Buckland's life spanned some of the most momentous changes in earth science, changes of which he was in great part the author. Not surprisingly for someone whose career combined science and religion at the highest level, he was an ardent follower of natural theology. One of Buckland's greatest

palaeontological discoveries concerned Kirkland Cave in Yorkshire where workmen found on the cave floor a huge assemblage
of broken, gnawed bones of hyaena, tiger, elephant, rhinoceros
and hippopotamus. All such creatures having long since van-
ished from Britain, the remains formed part of the evidence,
then being carefully gathered, that the earth had undergone a
change in climate, from warmer to colder. But what had happened in the cave during those ancient days ten thousand or so
years ago? The obvious explanation was that the bones represented carcasses washed into the cave by the Flood. Buckland
concluded instead that the cave, now some 100 feet above the
floor of a narrow valley, had originally been on the shore of an
ancient lake and that subsequently the water level had become
lower. The cave had been a long-occupied hyaena's den, and
here is where he showed his prowess as an experimenter as
well as observer. He took some cow bones and fed them to
the young hyaena he just happened to be keeping in the
garden. The tooth-marks left by the living hyaena exactly
matched the tooth-marks on the fossil bones. All this left the vast
intellectual edifice supporting the notion of a flood somewhat
weakened.

It was all brought crashing down by a brash young Swiss
zoologist-geologist named Louis Agassiz. In 1840 Agassiz
visited Britain, both to look at fossil fishes and to obtain more
evidence for a novel idea: that the surface features of northern
Europe had been shaped by water, all right, but by frozen water
– ice. If he was right, then a vast area of Europe had been
covered by a massive thickness of ice that, as could be observed
in modern glaciers, carved out valleys, moved immense volumes
of earth and rock around, and reshaped the entire landscape.
All the evidence that had seemed to support a flood, especially
the beds of redeposited sands and gravels, was material that
had been ground down and carried away by the inexorable

193

weight of these slow-moving seas and rivers of ice and by the torrents of water produced when the glaciers melted.

We now think that there were successive cycles of ice build-up and retreat on roughly a 100,000-year cycle. We are currently enjoying a relatively stable inter-glacial period that began only some 12,000 years ago. About 18,000 years ago, ice was near its peak over much of Europe where, as in central Greenland today, it was up to three miles thick, easily heavy enough to depress the earth's surface. For a huge range of scholars, a main prop of their accommodation to religion had been removed – there was no geological evidence for a worldwide flood. Buckland, for one, eagerly endorsed the new explanation. But for others, Charles Darwin among them, the glacial theory took some accepting. After he had returned from the *Beagle* expedition, he set off to Scotland to use the earthquake-uplift theory to explain the long-standing puzzle of the 'Parallel Roads of Glen Roy'. This phenomenon consists of step-like benches around the slopes of this and other valleys. Probably Darwin thought he could use this classic puzzle to make a splash in geological circles. Sure enough, he concluded with Charles Lyell that the 'roads' were a series of ancient sea levels – beaches – preserved on the sides of the valley as the land had been progressively lifted up, leaving the sea far behind. Darwin wrote up his ideas in a paper for the Geological Society of London. But he was wrong. In fact, like Buckland's cave assemblages, they reflect a lowering of the level of an inland lake damned by a glacier, not the raising up of the surrounding hills.

The idea of a Great Flood never seems to go away. In recent times, the new technologies of underwater exploration have produced something very interesting on the subject. When the glaciers began to retreat after the last glacial maximum, sea levels rose. Worldwide, they are on average some 300 feet higher than 18,000 years ago. For example, as recently as 10,000 to

Unfinished Business: Mountains and the Flood

12,000 years ago, Britain was still connected to continental Europe and the Thames was a tributary of the Rhine. The Mediterranean was also subject to this inexorable rise in sea level, the present region of the Bosphorus presenting a natural barrier, bottling up the rising water at its eastern end. Immediately beyond, the region of the present Black Sea was then occupied by a smallish inland lake with settlements on its shores, first at sea level, then below. As sea levels continued their relentless rise, there came a point when the Mediterranean could be held back no longer. Around 8,000 years ago, the barrier was breached and a vast volume of water burst through, swallowing up the land to the east and creating what is now the Black Sea. Underwater explorations have confirmed this, first finding beaches with freshwater shells from the older (smaller) lake bed, now 500 feet below the surface, and later finding apparent evidence of submerged occupation sites.[115]

To anyone on the ground the day the barrier was broached it must indeed have seemed as though the fountains of the deep had opened. Some caution needs to be expressed, however, before these new data can safely be interpreted as evidence for the great Biblical Flood. The dating may be wrong for Noah's Flood, which other sources such as the Babylonian *Epic of Gilgamesh* places much later.[116] And the ecology is wrong; if a freshwater lake had been replaced by salt water, the account in Genesis of the repopulation of the land after the Flood would have mentioned it. The bursting open of the Bosphorus into the Black Sea may have been real but the Flood, if it happened at all, is still more likely to have been a massive freshwater flood in Mesopotamia.

William Paley did not try to square the discoveries of cosmology and geology with the scriptures; the age of sacred theories and physico-theology was long dead in 1800. Instead, he closed his eyes to the things he could not deal with. He pronounced astronomy 'not the best medium through which to prove the agency of an intelligent Creator'. In geology, he ignored fossils, Hooke, Steno, Hutton, 'cycling', balanced theories, mountain-building, the age of the earth, and even the Flood. Obviously that strategy carried risks of its own, but it was (and is) a not unfamiliar style of argument. He just kept emphasising his main point: if he could show that the intricacies of nature could only have arisen from the Supreme Intelligence, everything else was moot.

CHAPTER NINE

This is Atheism

'We must not allow ourselves ever to reason without proper data, or to fabricate a system of apparent wisdom in the folly of a hypothetical delusion.'

James Hutton, *Theory of the Earth*, 1788

'Shall it turn us round to an opposite conclusion ... that no art or skill whatever has been concerned in the business? ... Can this be maintained without absurdity? Yet this is atheism.'

William Paley, *Natural Theology*, 1802

Although Charles Darwin created a theory of evolution driven by natural selection, only the latter part of the phrase was new to him. Evolution as a phenomenon, under various names like 'transmutation of species' and 'development theory', was on people's minds long before *On the Origin of Species* was published. Most new ideas, like Darwin's, tend to arrive not with a bang but a whisper. Orthodoxy has a thousand champions but something new, and quite often dangerous to conventional opinion, must make its own way, slowly and often precariously. As we have seen, throughout the eighteenth century, European intellectuals, whether from the scientific or the political side, cautiously explored the idea that life was not fixed at the instant of creation but had changed through time. Robert Hooke had entered this debate as early as 1694, observing that a reasonable inference from the evidence of the fossil record would be that 'there may have been ... Species in former Ages of the World that may not be in being at present, and many variations of those Species now, which may not have had a Being in former Times ... And to me it seems very absurd to conclude, that from the beginning things have continued in the same state that we now find them.'[117]

Interestingly enough, the first use of the word 'evolve' in its sense of organic change, occurred in 1826. The source was an anonymous article entitled 'Observations on the Nature and Importance of Geology' published in the Edinburgh New Philosophical Journal,[118] essentially forming a precis of the ideas of Jean Baptiste de Lamarck, first published in French in his *Philosophie Zoologique* of 1809.[119] The anonymous author states: '[Lamarck] admits on the one hand, the existence of the simplest infusory animals [creatures like amoebas and rotifers], on the other, the existence of the simplest worms, by means of spontaneous generation, that is, by an aggregation process of animal elements; and maintains, that all other animals, by the operation of external circumstances, are evolved from these.' Charles Darwin was a student in Edinburgh at the time, his teacher Robert Jameson was the editor of the journal, and it seems probable that either Jameson or Darwin's other Edinburgh mentor, Robert Grant, was the reticent author. Darwin must have been aware of this article, but in none of his writings does he reveal what, if anything, he then thought of such ideas except for allowing that he had read his grandfather Erasmus.

That concepts of evolution were alive and well (or in embryo, depending on your point of view) long before Charles Darwin was born is strikingly demonstrated by the fact that three of the atheistic heresies about which natural theologians were most worried in 1800 were evolutionary. Throughout Paley's book there is a consistent theme of resistance, not just to any revision of the notion of God the Creator, but specifically to theories of what we now call evolutionary change. Today we see Charles Darwin as the champion of evolutionary thought; in Paley's time the threats came from Charles's grandfather Erasmus and from a whole bevy of continental philosophers. William Paley tried ruthlessly to put down such hypotheses, not hesitating to call their authors atheists even if they had written in theist or deist faithfulness.

The most basic threat came from the world of Second Causes. For example, if motion is a fundamental property of matter and the cause of its other properties then, as Spinoza and Descartes had argued, motion was something inherent in matter and there would be no need of a cause outside of matter itself. The apparent purposiveness of nature might arise out of what Democritos called 'chance and necessity'. Causes of organised nature other than the intelligence of the Almighty, would exist in the lawful properties of nature. There would be no 'presiding Intelligence' behind living creatures and as the cause of life itself. Even the heretical theologian and philosopher, John Toland, for all his denial of the trappings of Christianity, could not go that far. Starting from the premise that matter could not have created itself, he demanded:

[Given that] God was able to create this matter ... is there likewise no necessity that he shou'd ever or rather always direct its Motions? Can the Formation of Animals and plants be accounted for from the Action, any more than the Extension of Matter? Or are you able to imagine that the Action and Reaction of Bodys, of all the Particles of Matter on one another, cou'd ever have the Contrivance to make any one of those admirable vegetable or animal Machines? All the jumbling of Atoms, all the Chances you can suppose for it, cou'd no more bring the Parts of the Universe into there present Order ... nor cause the

Organization of a Flower or a Fly, than you can imagine that by tumbling together the Letters of a Printer a million of times, they shou'd ever fall at last into such a Position, as to make the Aeneis of Virgil, or the Ilias of Homer.[120]

This last simile reminds us of the familiar anti-Darwinian jibe that a thousand monkeys hitting randomly at typewriters could not produce a single line of Shakespeare. The argumentative Reverend Ralph Cudworth had used the same analogy: '[It is] no more possible that Dead and Formless Matter fortuitously moved, should at length be Taught and necessitated by itself . . . than a dozen or more Persons, unskilled in Musick, and striking the strings as it Happened, should at length . . . fall into Exquisite Harmony.' And of course, its origin is classical. Cicero had written, two thousand years earlier,

> Must I not marvel that there should be anyone who can persuade himself that there are certain solid and indivisible particles of matter borne along by the force of gravity, and that the fortuitous collision of these particles produces this elaborate and beautiful world? I cannot understand why he who considers it possible for this to have occurred should not also think that, if a countless number of copies of the one-and-twenty letters of the alphabet, made of gold or what you will, were thrown together in some receptacle and then shaken out on to the ground, it would be possible that they should produce the Annales of Ennius . . . I doubt whether they could possibly succeed in producing a single verse.[121]

Atheism in the form of evolutionary ideas opened the door for the discovery and investigation of second-cause mechanisms flowing from the properties of material nature. Everything was focussed on two interdependent questions: how had life arisen on earth, and what were the causes of subsequent changes in life on earth? Theories of evolution in various early guises were philosophically attractive to many because they combined aspects of the 'atomick' model with a mechanism of change grounded in the empirically observed facts of nature. Evolution was always more than a purely scientific notion, however, and both before and after Charles Darwin its consequences were felt in a larger political context than the narrow debate over Genesis.

Evolution basically refers to change (The Latin *evolvere* means to develop or unfold). Whether as an actual mechanism or a metaphor, evolution was an attractive notion to political theorists because of all the liberty, equality and self-determination that was bound up in the single word 'change' when it was uncoupled from and ranged against predestination. If humans, just like animals and plants, are the product of change, then we in turn *may* change. It can be extended from mere physical properties, such as the giraffe developing a long neck, to the intellectual and social world. In a world of change, we may even control our own destinies, and if our grandchildren may accomplish more than us, whole societies may change; that is a true freedom. Equally, evolution is the natural enemy of those who would prefer the older certainties and the status quo. Those who would see rank and station, power and privilege passed down unchanged from generation to generation within God-specified lineages were alarmed by theories that fitted only too well with the rise of a new middle class and movements for democracy. 'Change' was a more fitting watchword for the French Revolution than for Britain's Restoration of the Monarchy, a hundred years before.

Because 'evolution' has itself evolved, it includes several different concepts. As a result, a lot of nonsense is talked about it, both from those who abhor the notion and those whose research demonstrates it. The modern core sense of the term is that life has changed over time, and that the course of change has been one of diversification so that all extinct and living organisms are related by descent from common ancestors. The fact that life on earth (like the earth itself) has changed over time, in consistent patterns, is quite independent of any theory about the causes of change and is constantly confirmed in the following way. The fossil record shows that, over more than 2 billion years of earth history (one can refine the exact span of time without changing the fact), life on earth has changed, always in the direction of simple to complex (although some secondary reversals, like simplification of body form in parasites, must be granted). Life has, as it were, explored in a multitude of directions while constantly diversifying; many, perhaps most, of these experiments were dead ends. Among the vertebrates, first there were fish, then amphibians and reptiles (and birds), then mammals. Among mammals, first there were shrew-like creatures, later elephants, cows, monkeys and, most recently, humans. Some kinds of creatures, like the coelacanth fish, became protected from, or immune to, change in ways that we still do not fully understand, and persisted as 'living fossils'. The whole panoply of life on earth forms a series of patterns within patterns, all changing over time. And yet it is also all very simple. A trout, a shark, a goldfish, a tuna, are all different ways of being the same thing – a fish. Fish, frogs, birds and humans are all different ways of being an animal. Under the skin, our DNA tells us, worms, flies and vertebrates are all cousins.

The fossil record is the factual, historical evidence of these relationships. The facts of the fossil record are tested every day; more details about the patterns may be added as our knowledge increases, changes may be made to various elements, but the overall consistency in the pattern remains. It is an empirical fact that in Devonian rocks there are no dinosaurs. There are no mammals in the Silurian, no humans in the Mesozoic (when the dinosaurs lived), and so on. If one were to look closely at a small subsection of the fossil record, the same consistency of temporal sequence would be found at a finer scale. At every hour of every day, somewhere in the world, a palaeontologist is busy hacking fossils out of the rock and explicitly testing the temporal patterns of change. The fact of evolution as a process of change over time is constantly tested, and has never failed.

The pattern or sets of patterns in the history of life revealed by fossils are the logical extension of what John Ray and Linnaeus, among others, had discovered about the diversity of *living* forms on earth, which was simply a snap-shot, as it were, of change in process. From Aristotle onwards, we have known that the so-called 'natural groups' of organisms can be arranged in hierarchical systems that tend to fall out along lines going, yet again, from the more simple to the more complex. But science always wants to find out not only what is, but also the whys and the wherefores. Before the final demise of the concept of spontaneous generation, one way to explain the pattern of life was to suggest that the progressively more complex lineages of animals and plants had derived from a sucession of discrete generation-events over time. After the possibility of a continuously acting spontaneous generation had been decisively confounded, natural philosophers were left either to subsume all the groups of animals and plants under one initial act of creation (after which they had remained unchanged), or to find ways in which the earliest, simplest forms of life might have given rise, through Second Causes, to the later ones. But the problem remained: by what process, what materially and scientifically definable mechanisms, could life have arisen and such changes – any changes – have been produced? (One interesting modern development has been to postulate that life on earth arose from (multiple) invasions of life from space, but that does not explain where and how *that* life originated.)

In order to make the case for natural theology, William Paley had to take up arms against evolution in all its forms; that is why it is so interesting to know that Charles Darwin read and absorbed Paley's arguments as early as 1831. Paley's first battle was with all the Cartesian randomness that evolutionary change implied:

> The Philosopher beholds with astonishment the production of things around him. Unconscious particles of matter take their stations, and severally range themselves in an order so as to become collectively plants and animals, i.e. organised bodies, with parts bearing strict and evident relation to one another, and to the utility of the whole; and it should seem, that these particles could not move in any other way than as they do, for they testify not the smallest sign of choice, or liberty, or discretion. (There may be plastic natures, particularly intelligent beings, guiding these motions in each case: or they may be the result of trains of mechanical dispositions, fixed beforehand by an intelligent appointment, and kept in action by a power at the centre,) 'but in either case there must be intelligence'.
>
> Others [Paley means Descartes] have chosen to refer every thing to a principle of order in nature. [But] order itself is only the adaptation of means to an end: a principle of order, therefore, can only signify the mind and intention which so adapts them ... a principle of order, acting blindly and without choice, is negatived by the observation, that order is not universal; which it would be, if it issued from a constant and necessary principle.

A principle acts blindly rather than purposefully. To Paley, that was impossible: '*What is sometime called nature, sometimes called a principle: which terms, in the mouth of those who use them philosophically, seem to be intended, to admit and to express an efficacy, but to exclude and to deny a personal agent.'*

But materialist philosophers (particularly those in mid-eighteenth-century France who had shed their religious convictions or who were untrammelled by them) had already gone far beyond mere vague imaginings of the chance motions of atoms. They had begun to consider mechanisms that would describe the organisation of matter and that did not depend on chance but grew out of natural laws. Many, like Hume, looked for clues in the phenomena of 'generation' (reproduction and development) and 'vegetation' (growth) – for example, 'plastick virtue'. With respect to fossils, Martin Lister had thought that 'formed stones' could not simply arise out of nothing but that there must be some property of the earth that 'caused' them. None other than Dr Ralph Cudworth (see Chapter 3) was among those who took the notion of quasi-Newtonian recycling of matter in nature – life and death, ashes to ashes, dust to dust – and looked to see whether there was some such property of the 'soil' that directed the formation of new life. The same idea had been proposed under astrological principles by the sixteenth century doctor and botanist Cesalpinus and his followers, whom Richard Bentley in the first of the Boyle Lectures dismissed as 'Astrological undertakers, that would raise Men like Vegetables out of some fat and slimy soil, well digested by the heat of the Sun, and impregnated with the influence of the Stars upon some remarkable and periodical conjunctions'.[122] (Atheism here consisted of substituting some 'plastick' faculty of nature for the hand of God.)

In France, Charles Bonnet (1720–1793) developed Aristotle's idea that the different kinds of organisms represent a sort of continuum. What appears to us as an ascending order of complexity, from the tiniest microbe to man, is a system that he called the *scala naturae*. Bonnet thought that God originally created a whole series of 'germs', contained one within another, each capable of turning into a different kind of organism. Over the history of the world, successive germs developed, steadily increasing in 'perfection', mounting the ladder of life towards man. And man contains germs that will eventually develop into an even higher organism, approaching but never to reach the perfection of God. This *scala naturae* thus implied a notion of progress – at least in the mind of God – with man, his more recent and greatest creation, following his lesser works. In the end, however, all interpretations of causality in nature reduce to two: either the orderliness of any pattern is the product of God's mind (whatever we might mean by 'mind' when referring to the deity) or else it derives from the orderliness of some independent *process* and *property* of material nature.

Another Frenchman, Georges Louis Leclerc, Compte de Buffon (1707–1788), one of the dominant scientists of the eighteenth century, asked the key question about the 'causes' of life as demonstrated in the mechanisms of reproduction and growth: when animals and plants create their own bodies out of simple raw materials, how do the various atoms, molecules, or 'corpuscles' know where to go? What guides each component to its right place, one to form part of a muscle, another a bone? When a bone breaks, how is new bone made in the gap between the broken ends? What organises all this? Buffon's answer was to propose a kind of internal blueprint – the *moule interieur* ('internal mould'). Like Paley, he argued from an analogy, and once again it was the analogy of death and rebirth, the cycle of the farm and the garden: life to compost to life. Living systems break down after death into their non-living constituents. Instead of an organising principle in the soil and rocks, Buffon proposed that for each species there was a biological internal mould and the nutrients that the animal or plant consumed somehow passed through this mould and were processed into the right form. This was exactly the sort of mechanistic materialism that Paley knew contained the ultimate threat to the religious view of the creator's role, even if such a system would normally operate in a stabilising way, producing carbon copies of the original. Only too obviously, if for any reason the mould should change, for example under pressure from the environment, the result would be some new combination of properties – and the emergence of a new species. (Not only does all this smack of 'Plastick Vertue', it can also be seen as a nice metaphor for the blueprints that we now know DNA to be).

Obviously the word 'mould' was only used in a metaphorical sense, but that was a weakness for Buffon: once again, the ideas far outstripped the data. It was hard to see what evidence there was for such internal moulds and how they would work. Paley, apparently unaware of the similarity of an internal mould to his 'system of laths and files', seized on this weakness: 'Does any reader annex a meaning to the expression "internal mould" ... Ought it then to be said, that, though we have little notion of an internal mould, we have not much more of a designing mind?' Nonetheless, it is obviously true that at each cycle of birth, new life appears in apparent mimicry of the origins of the first life. In principle, there had to be a mechanism of some kind. As Paley himself wrote in the final pages of his book, 'In the ordinary derivation of plants and animals from one another, a particle, in many cases, minuter than all assignable, all conceivable dimension, – an aura, an effluvium, an infinitesimal, – determines the organisation of a future body: does no less than to fix, whether that which is about to be produced shall be a vegetable, a merely sentient, or a rational being.' However, the secrets of reproduction and morphogenesis were only being revealed very slowly to contemporary natural philosophers, because they involved single cells (sperm and ova) and processes at a microscopic level. No one could postulate the mechanism by which a hydra grew a new head. The other difficulty was that people were incapable of letting go of their preconceptions long enough to examine life dispassionately. When they began to do so, they could take what had been seen as God's creative power and assign it to Second Causes; of these, 'internal moulds' were only the beginning of a flood of discoveries of biological mechanisms.

Behind 'plastick vertue' and 'internal moulds' there was a hidden truth, the explanatory power of which David Hume had brilliantly articulated a hundred years earlier on purely theoretical principles and summed up in the general concept of 'generation'. 'Generation' offered great potential in cracking the problem of the mechanisms and causes of change. In his *Dialogues*, Hume had written: 'Judging from our limited and imperfect experience, generation has some privileges above reasons [i.e., design]; for we see every day the latter to arise from the former, never the former from the latter ... reason, instinct, generation, vegetation, are similar to each other, and the causes of similar effects. What number of other principles may we naturally suppose in the immense extent and variety of the universe ... ?' If the processes of reproduction are constantly at work creating new life, could not 'generation' also create change?

Paley, so much of whose *Natural Theology* was committed to refuting Hume, understood the dangers but could only reply in rhetorical terms:

What principle of generation ... ? The minds of most men are fond of what they call a *principle*, and of the appearance of simplicity in accounting for phenomena. Yet this principle, this simplicity is sometime nothing more than in the name ... The power, in organized bodies, of producing bodies like themselves, is one of these principles. Give a philosopher this, and he can get on. But he does not reflect, what is this principle . . . Yet, because the whole of this complicated action is wrapped up in a single term, generation, we are to set it down as an elementary principle: and to suppose, that, when we have resolved the things which we see into this principle, we have sufficiently accounted for their origin, without the necessity of a designing, intelligent, Creator . . . The opinion, which would consider 'generation' as a *principle* in nature . . . is confuted, in my judgement, not only by every mark of contrivance . . . but also by the further consideration, that things generated possess a clear relation to things *not* generated.

Paley argued that generation produces an animal with lungs, but it does not produce the air: 'yet their properties correspond'. Similarly the eye 'corresponds' with light, and the ear with '*undulations* of the air'. For Paley, therefore, adaptation of the living to the non-living proved that something outside the living system of 'generation' or 'internal moulds', and independent of the nature of non-living matter, was responsible for the design of life. He challenged his reader: 'Is it possible to believe that the eye was formed without any regard to vision, that it was the animal itself which found out that . . . it would serve to see with?' (Here again is Hume's question: which came first, the part or the use?)

The very language that the natural theologians used betrays the dangers lurking in the philosophical undergrowth. For example, the great Unitarian preacher/scientist/social reformer Joseph Priestley was equally dismissive of these atheistic ideas, writing contemptuously in a review of Hume's *Dialogue Concerning Natural Religion*:

With respect to Mr Hume's metaphysical writings in general, my opinion is, that, on the whole, the world is very little the wiser for them . . . Had any friend of religion advanced an idea so completely absurd as this, what would not Mr Hume have said to turn it into ridicule. With just as much probability he might have said that Glasgow grew from a seed yielded by Edinburgh, or that London and Edinburgh, marrying, by natural generation, produced York, which lies between them.[123]

Last and by far the most dangerous of the eighteenth-century 'atheists', and much closer to home (in every sense) than Buffon and his internal moulds, was that infinitely complex character Erasmus Darwin, who first started toying with evolutionary ideas as early as 1769 as a result of thinking about the meaning of fossils. It became clear to him that fossils, and especially those of extinct creatures, indicated a continuity of life that was best explained by a process of change, and if the pattern of life had changed over time, all living creatures might, as ancient Greek philosophers had proposed, have descended from some single ancestor. Prudent friends persuaded Darwin not to publish his ideas, and it was some twenty-five years before he dared to include his revolutionary theory in the first volume of an enormous work called *Zoonomia; or, the Laws of Organic Life*.[124]

This rambling compendium of essays is largely directed towards a theory of disease. But here, in the thirty-ninth of forty essays, Darwin finally expounds the theory he has been confiding to his friends for so long. All life is descended from a single source, which he envisages as a microscopic 'filament'. There is then a continuity between this original filament and the microscopic filaments which are the origins of the embryos of all creatures: 'produced from a similar living filament. In some this filament in its advance to maturity has acquired hands and fingers, with a fine sense of touch, as in mankind. In others it has acquired claws or talons, as in tygers and eagles. In others,

toes with an intervening web, or membrane, as in seals or geese.'
Erasmus Darwin thus tried to develop both a logic and a mechanism out of the principles of generation, about which Hume
and others had speculated. As generation provided the continuity among all living creatures, their fundamental relationship
was the result of (was caused by) modification of the filament.
Of course the filament was speculative, even metaphorical, but
Darwin, recognising that many traits of living creatures are
inherited, had gone a little further along the long road towards
a theory of genetics and evolution.

In one of his most famous passages Erasmus Darwin asked:

> Would it be too bold to imagine, that in the great length of time
> since the earth began to exist, perhaps millions of years before
> the commencement of the history of mankind . . . that all warm-
> blooded animals have arisen from one living filament, which the
> GREAT FIRST CAUSE endued with animality, with the power
> of acquiring new parts, attended with new propensities . . . and
> thus possessing the faculty of continuing to improve by its own
> inherent activity, and of delivering down those improvements
> by generation to its posterity, world without end.

There is an awful lot contained in this one statement. Here he
piously (as a deist) grants the First Cause, but ends with a rather
ironic-sounding benediction. Elsewhere, he even proposed a
chemical origin of life – 'without cause'.

Darwin also knew that the world must be very old, even
millions of years old, and that change had proceeded (and
accumulated) for aeons before humans appeared. He averred
the continuity of life from his microscopic filament source and
then produced the mechanism by which change might have been
produced.

It appears that all animals have a similar origin, viz. from a single living filament; and that the differences of their forms and qualities has arisen only from the different irritabilities and sensibilities, or voluntarities, or associabilities, of this original living filament; and perhaps in some degree from the different forms of the particles of the fluids, by which it has been first stimulated into activity. And that from hence, as Linnaeus has conjectured in respect to the vegetable world, it is not impossible, but the great variety of species of animals, which now tenant the earth, may have had their origin from the mixture of a few natural orders.

Change is thus produced through the reproductive processes, where 'fibrils with formative appetencies, and molecules with formative aptitudes or propensities ... combine, and form the primary parts of the new organization of an embryon'.

The creation of new life-forms, then, depended on an array of 'efficient and inert' causal factors acting upon these appetencies, aptitudes and propensities. Among these were the drives for reproduction (lust), food (hunger) and 'security' (as revealed in new adaptations for survival, such as 'fleetness of foot', or shells, or mimicry). Thus 'some animals have acquired wings instead of legs, as the smaller birds, for the purpose of escapes'. The weakness in his theory (although for a while it gave it a popular acceptance) was that, while built upon the ideas of mutation, hybridisation and pressures of the changing external environment, it depended largely on what we now call the inheritance of acquired characteristics. For example, if the blacksmith developed brawny arms, his children would have brawny arms.

Paley seized upon the key philosophical danger in Erasmus Darwin's ideas: 'in the formation of plants and animals, in the structure and use of their parts, it does away with final causes ... Give our philosopher these appetencies ... and, if he is to be believed, he could replenish the world with all the vegetable and animal productions which we at present see in it.' And on the more practical side: 'The solution does not apply to the parts of animals, which have little in them of motion. If we could suppose joints and muscles to be gradually formed by action and exercise, what action of exercise could form a skull, or fill it with brains ... [and] ... in the last place, what do these appetencies mean when applied to plants? ... A solution is wanted to one, as well as the other.' In fact, Erasmus Darwin had professed himself very comfortable with the idea of God as the First and the Final Cause. But it was the same old problem: good ideas can easily be dismissed if they are not supported by understandable, experimentally verifiable mechanisms (although the same critics have no trouble accepting an equally non-verifiable idea of an intelligent, designing, creative mind).

Just after Paley had completed *Natural Theology*, Jean Baptiste de Lamarck (1744–1829, Botanist of the King at the Jardin du Roi in Paris and, after the Revolution, Professor of Zoology at the Museum National d'Histoire Naturelle) assembled a different, arguably more coherent, theory of evolution. As a straightforward piece of natural philosophy, Lamarck's *Philosophie Zoologique* had none of the literary pretensions and limitations of Erasmus Darwin's works but contained the same core idea, although with major differences in the detail. To start with, Lamarck proposed that the different lines of organisms had not been created all at once, but rather progressively over time, by repeated spontaneous generation. His was the last serious attempt to invoke spontaneous generation. After each creation event, the organisms in each lineage rather mysteriously progressed up a version of the old *scala naturae*. Mammals had at earlier times been fish. However, all fish were not necessarily created before mammals; they could instead belong to newer lineages that had not yet progressed very far up the ladder of life. A neat corollary of this was that

Jean Baptiste de Lamarck

there was no such thing as extinction. Ichthyosaurs might have been replaced by whales and dolphins but at some time in the future a given lineage would be in the right position on the ladder to produce them again. As he could not specify the mechanisms by which such large-scale progressions of lineages along this escalator of life might actually happen, Larmarck's theory was no less mystical than others.[125] But in invoking an end-directed (teleological) process, Lamarck was keeping faith both with classical philosophers and with Christians. In one other respect Lamarck hit a responsive chord, taking over Erasmus Darwin's ideas about the inheritance of acquired characteristics, and giving his famous example of successive generations of giraffes straining to reach the prime shoots at the tops of trees, thereby developing into a longer-necked species. This mechanism would explain deviations from the otherwise inexorable progress up the ladder and therefore create the impression of diversification. It was only a small piece of Lamarck's theory but he has unfairly been saddled with it, in popular estimation, as his principal legacy.

214

Lamarck's theory may have been no better than Erasmus Darwin's but there is no doubt that he achieved, in the early nineteenth century, a far broader and more sympathetic following. Darwin had published his ideas when philosophies of change were being discredited in England by the revolutionary events of France; Lamarck's ideas were planted in the more fertile, if no less revolutionary, soil of the 1820s to 1850s. He embodied two centrally important movements: progress, and change that is individually driven (by will). In fact, popular support for Lamarck continued to be so great that Charles Darwin even included specific allusions and references to Lamarckism in later editions of *On the Origin of Species*.

Buffon, Erasmus Darwin and Lamarck all had in common two principles: first, a sense of a causal generative process, normative in nature, and second, the possibility that change (diversification, transmutation) was the result of deviation from the norm, driven by chance or circumstance. Paley's lathes and files, and even the various theories of 'plastick vertue' espoused by people like Lister, are all attempts to accommodate the first principle but all range dangerously near to being capable of extension to the second sense also. It seems unlikely that Paley was unaware of this. There is therefore something tantalising and almost cavalier in his dismissal of Buffon and Erasmus Darwin, particularly in the challenges he throws out. If Buffon's ideas are correct, he asks, 'if the case were as here presented, new combinations ought to be perpetually taking place: new plants and animals, or organised bodies which were neither, ought to be starting up before our eyes every day . . . Is it known [that] any desert has been thus repeopled?' And, right at the beginning of *Natural Theology*, he had complained: 'Multiples of conformation, both of vegetables and animals, may be conceived capable of existence and succession, which do not yet exist. Upon the supposition here stated, we should see unicorns

215

and mermaids . . . nations of beings without nails upon their fingers . . . with more or fewer fingers and toes than ten.' This is almost a recipe for a research programme. One wonders how many contrary-minded readers of Paley, in his own lifetime, thoughtfully looked around themselves – especially at the fossil record – or made a mental note to think about these questions. Robert Chambers, the most influential nineteenth-century writer on evolution before Charles Darwin, was, for example, one of the rare humans with six fingers. And one reader in particular must have reacted to the sting of Paley's sarcastic diatribes against Erasmus Darwin. It would be surprising indeed if the young Charles Darwin had not himself made a mental note to see if he might discover some way of vindicating his grandfather. Whether he explicitly had Paley in mind as he started to think about the meaning of the cacti, finches and tortoises that inhabited the 'deserts' of the Galapagos Islands, or whether he reinvented the questions for himself, we will never know. However, where Paley asked to be shown where new species had arisen, Darwin eventually showed it.

The principal drawback to theories of transmutation like those of Buffon and Erasmus Darwin is that while, as Hume and so many others had pointed out, there was a logical connection between the 'creation' of a living embryo (whether of a human or a worm) from its parents and the creation of new species, no one had the slightest idea about the essential properties of reproduction and how the complexity of an embryo could arise from the apparent simplicity of a single cell. The various 'germs' and 'principles' in the speculative (philosophical) literature were simply metaphors and had as yet no empirical basis in science. In fact, very little of the internal workings of living organisms – the liver, kidney, lung or heart, let alone the brain – was known in a mechanistic sense. This gave natural theology a position of relative security from which to deny such theories and to substitute for them a vitalist mystery. It would be another 150 years or more before physiology could be explained in terms of biochemistry, and chemistry in terms of physics. Obviously, a theory of evolution that was not wrapped in the mysteries of 'generation' (reproduction and genetics), but was instead expressed in terms of simple common-sense phenomena, would stand a better chance of acceptance. Charles Darwin created just such a theory with his 'natural selection'. Once *On the Origin of Species* had been published, the arguments of Buffon, Erasmus Darwin, Hutton, Paley, Priestley, and all the others were pushed to the sidelines.

The twin premises of modern evolutionary thinking are that life arose once (by natural processes and out of non-life) and that all life proceeds from existing life. If many forms have arisen from one, then implicitly evolution is a process of diversification and branching. This statement could be rejected if it could be shown that life has arisen more than once (this can be tested by examining whether all organisms have related DNA patterns – which they do). Therefore when Samuel Clarke in *Demonstration of the Being and Attributes of God* (1705) asked: 'What absolute Necessity [is there] for just such a Number of Species of Animals or Plants?', the modern biologist answers that diversification flows inevitably from the nature of the process. With his concepts of natural selection and sexual selection, Charles Darwin provided a causal mechanism to account for that process and for the crucial phenomena of change, relationship and diversification, which was corroborated by an extraordinary array of observational and experimental data.[126] Darwin's natural selection may not be the only causal agent of evolutionary change (certainly random effects must play a part, for example), but it is the basis of the most complete theory we currently have and seems sufficiently well supported that it will only be complemented by other theories, rather than replaced.

The full origins, intellectual debts and sources of Charles Darwin's theory of natural selection may never be known. During the *Beagle* voyage he had come to recognise that the earth was subject to enormous upheavals. His study of the fossils of South America showed that the extinct species were so closely similar to living species as to suggest that one had replaced the other over time. His chronicling of the natural history of South America revealed that many living species were also distributed geographically in such a way that they seemed to replace each other in space as well – one species to the north, a different but very similar one taking over to the south. During and after the voyage, when he came to look over the natural history of the Galapagos Islands, he saw the truth of what a government official had told him: each island had its own version of the same animal or plant. There was a different kind of (very similar) tortoise and cactus, and groups of different ground finches, on each island. At first Darwin thought the finches were all minor variants (races) of a finch on the mainland, but when the ornithologist John Gould studied them, he told Darwin that each was a separate species. Therefore, Darwin concluded, they must have arisen on the islands by simple genealogical descent from a common parent. Similarly, the unique Galapagos tortoises and iguanas had descended from some mainland species that had somehow survived being swept out to sea, their descendants slowly evolving to new species in situ.

By 1838, Darwin had worked out the mechanism – natural selection – that could be the cause of such change over a long period of time. The essence of Darwin's theory is remarkably simple, being the logical conclusion from four incontrovertible facts. The first of these is 'variation'. In any group of living organisms – whether the Macaroni penguins on South Georgia or the human passengers in a commuter bus – the individuals differ. Penguins may all look the same to us, but to other penguins they do not. Some are slightly taller or thinner, each may have a different level of resistance to a particular disease. Humans, similarly, not only differ notably in height, weight and eye colour, but also in everything from temperament and metabolism to susceptibility to alcoholism and breast cancer. Secondly, and crucially, many of the features by which we differ are genetically caused and therefore inherited. Mutation constantly produces new variants – if this were not so, we would all look exactly the same except for those differences caused by diet, exercise and the physical environment. Only genetically based differences can in turn be inherited.

The third fact is equally incontrovertible. Mathematically, if population numbers of any and all species are to remain stable – which on the whole they tend to do – then of all the offspring that are ever produced over one generation, only two would be needed, on average, to contribute to the next generation. But all living organisms have the capacity (super-fecundity) to produce many more offspring than that over their reproductive lifetimes. For example, a single female cod may lay several million eggs per year for twenty years; they cannot all survive. At the other end of the spectrum, Darwin noted, 'The elephant is reckoned to be the slowest breeder of all known animals . . . [yet] at the end of the fifth century there would be alive fifteen million elephants descended from the first pair' if not naturally checked.

The fourth fact is self-evident: a whole variety of factors prevent this overpopulation from happening. Some offspring are sickly from birth, many are lost in the normal strife of a deeply competitive world in which most organisms are the food for some other organism. Others simply succumb to disease, to the elements and to accident long before they can reproduce. The vicissitudes of life therefore form a powerful filter, allowing only certain individuals through. These are the ones that are either extremely lucky or are best suited to deal with the problems that life throws at them: the elephants with the most mobile trunks – not too short and not too long – the flattest fleas, the smartest primates, and so on. (Of course, it is the whole organism that is constantly tested, not just one attribute.) In practice, then, some lineages will contribute more individuals to the following generations than others. This is natural selection and, as will be developed in the following two chapters, it owes a surprising debt to Paley's *Natural Theology*.

Natural selection operates inexorably, all the time. First and foremost it acts as a weeding-out process maintaining the status quo at some optimal level.[127] It serves to match the properties of the population and species to the prevailing conditions by selecting which individuals will or will not reproduce and contribute to the future stock. Otherwise, the characteristics of a population might drift out of 'fitness' by chance. Individuals do not evolve, only the sum properties of the higher group. If, over long periods of time, conditions on earth change – for example, the moving continents produce huge changes in ocean and atmospheric patterns as well as new seas and new lands – or if, by some chance, a population becomes isolated in some new place (like the Galapagos), the filter will automatically be reconfigured to allow through a different subset of the individually variant organisms from before. In this case, the same unyielding process of natural selection produces change. Humans create an analogue of this in their artificial selection of breeds of dogs, varieties of fruit and strains of wheat. (Equally, were it not for the benefits of modern medicine, humans with asthma, for example, or very bad eyesight, would be at a disadvantage compared to the rest of the population and in nature one would expect selection steadily to weed out lineages in which these conditions appeared.) Natural selection, acting over far longer time frames than our artificial selections, produces new varieties, new species and (over aeons) completely different kinds of organisms, *and removes the old versions*. Obviously this is a system in which, as Lucretius and Hume had foretold, the parts (the variations in form and behaviour) come into being before the use (survival both on a daily basis and into future generations) after all.

A key additional element of the pattern of change is that, under selection, a population or species may become divided into two or more descendent lines. Just as all modern breeds of dog are derived from a common wolf- or jackal-like ancestor, so all living woodpeckers are descended from some ancestral woodpecker and all humans from a single population of an ancestral ape. Natural selection is therefore a mechanism that is capable of *causing* the three principal elements of evolution: change over time, descent from common ancestors and diversification.[128] Because natural selection is an ongoing process, resulting from the essential properties of variation and overproduction, we can examine the living world for evidences of its operation. We can document overproduction, we can document variation, we can document cases of natural selection. All around us are cases of species *in flagrante delicto* evolving daughter species, or becoming extinct.

Thomas Jefferson once remarked that nothing was more inevitable than an idea whose time has come. Alfred Russell Wallace came up with the theory of natural selection a little after Darwin. But Darwin was not the first to develop the elements of such a theory. Patrick Matthew, a gentleman-farmer and naturalist, outlined a decent version of a theory of variation and natural selection, based on the analogy of artificial selection of plants and animals by breeders, in his 1831 book *Naval Timbers*. He even used the term 'natural process of selection'.[129]

Furthermore, as early as 1794 none other than the polymath-geologist James Hutton had mused on this subject.

> Now, this will be evident, when we consider, that if an organised body is not in the situation and circumstances best adapted to its sustenance and propagation, then ... those which depart most from the best adapted constitution, will be the most liable to perish, while, on the other hand, those ... which most approach the best constitution ... will be best adapted to continue, in preserving themselves and multiplying the individuals of their race ... Now, if those organised bodies shall thus multiply, in varying conditions according to particular circumstances ... we might expect to see ... a variety in the species of things which we might term a race.[130]

Hutton was not talking about the inheritance of acquired characters but a version of selection based on 'a varying power in the seminal production of organised bodies'. He used it to explain the origins of races, not species. However, neither Hutton nor Matthew put the whole theory together.

The strength of Darwin's theory (then and now) is that it did not rely on arcane metaphors or require the postulation of a mechanism that could not be expressed or understood by the layman. It was based on hard, simple realities observable all around us and regularly practised in the farmyard. Even so, science always continues to question and refine even its most central tenets, and Darwin did not for one moment imagine that he had written the last word on evolution. He could say nothing, for instance, about the deeper mysteries of generation – but then his theory did not depend on them. He continued to modify his ideas during his lifetime. What we call 'evolutionary theory' today, as the complexities of genetics and the genetic code are finally being deciphered, is vastly more refined than in 1859 and yet at the same time remarkably true to its origins.

CHAPTER TEN

Gosse's Dilemma and Adam's Navel

' 'Tis a dangerous thing to ingage the authority of Scripture in disputes about the Natural World, in opposition to Reason, lest Time, which brings all things to light, should discover that to be false which we had made Scripture to assert.'

Thomas Burnet, *Archaelogiae Philosophicae*, 1692

By the mid-nineteenth century, there were really only three ways in which natural theologians could deal with the growing evidence that the earth was very old, that it was recycling inexorably beneath their feet, and that life on earth had constantly changed over millions of years. They could ignore it, they could accommodate it to the biblical accounts of history by more or less denying the literal truth of Genesis, or they could explain it all away. The later natural theologians largely ignored it. The sacred theorists tried unsuccessfully to reconcile geology with the Bible. And one man above all others tried to explain it away. He was Philip Henry Gosse (1810–1888), a writer on natural history whose books caught the imagination of generations of Victorians and whose life became a tortured tale of religion contesting with science.[131] He lived and wrote long after Paley had died and there are still some who adhere to his conclusions.

Gosse's dilemma was that of all natural theologians, especially after the publication in 1844 of an anonymously authored, thrillingly dangerous, and wildly successful book on evolution.

223

This book went into many editions and earned first notoriety then fame and fortune for its author, who was not Charles Darwin but Robert Chambers of the Edinburgh publishing family. The book's title, with an allusion to James Hutton that nobody could miss, was *Vestiges of Creation*.[132] Chambers' theory was largely derived from Lamarck's which, like Erasmus Darwin's, depended upon organisms being subject to change as a direct result of environmental pressures and exigencies. Chambers probably set Charles Darwin back fifteen years – much to the benefit of all. In many ways he blazed the trail that Darwin could more cautiously follow with an even more convincing theory in hand. Darwin must have realised, with the example of Chambers in front of him (and approval of the political left and censure from both the religious and scientific right) that he would have to ensure his theory would have a better reception.

Gosse knew that various versions of what we now call evolution had been around for more than a hundred years. By the mid-1850s, most scientists in Britain knew which way the wind was blowing. Darwin had been hard at work in private since 1842, preparing the ground for his idea of natural selection, and knowing how popular a scientist Gosse was, he tried to enlist him to support his theory.[133] Darwin's self-designated 'bull dogs', including Thomas Huxley, were steadily persuading the sceptics – Huxley had been lecturing formally on an evolutionary relationship between men and apes as early as 1858.[134] This growing evolutionary movement offered a new way of explaining the evidence of organic changes, but only at the expense of much accepted religious belief. It threatened to change radically the whole frame of intellectual reference and to produce a new explanation of cause. For a huge number of theologians, clerics, philosophers and ordinary people, evolution was changing the metaphysical balance of power. Among those who felt this most keenly was Gosse.

Philip Henry Gosse

One's heart has to ache for Gosse, one of the most sympathetic characters of the evolutionary saga, a man weighed down by the burdens of fundamentalist Christianity and at the same time a brilliant naturalist. He occupies a treasured and honoured place in British intellectual history, writing about science and about his travels in Canada and Jamaica. He was the first to introduce to a popular audience the life of the seashore, the fragile world of exquisite beauty and strength that lies just a few inches beneath the surface of the sea and in the rocky pools of the coast. Before Gosse, all this was largely unseen. Gosse single-handedly created marine biology and home aquaria, and became one of the great chroniclers of the intricate worlds revealed by the microscope.

In every sense, Gosse was a product of his environment; he found his scientific inspiration in the marine biology of Devon and his religious life in membership of the Plymouth Brethren, a dissenting sect that from its origins around 1830 quickly split to form at least six different variants, all around the core belief that the only priesthood a church requires is that of its members. For the Plymouth Brethren, then, there was no need of a Holy Ghost to anoint special vicars of God on earth. In 1857, with his first wife dead of tuberculosis and his family half-starving – literally from his difficulty in eking out a living as a writer and lecturer, and figuratively from the self-denying rigours of the lifestyle of the Brethren – they moved from London to be closer to his beloved seashore and to fellow believers. His immortal place in intellectual history was doubly secured by his son Edmund Gosse in the pitiless yet endearing *Father and Son*, in which Gosse's life during those years as a devout member of the Brethren is dissected for its naive intolerance and carefully measured love.[135]

Once Lamarck and Chambers had made it possible (even necessary) to take evolution seriously, and after his meeting with Charles Darwin had shown how powerful was the extent of the challenge to his fundamentalist beliefs, Gosse felt called to respond; as a Plymouth Brother and as a scientist, it was his responsibility, just as it had been Paley's and before Paley John Ray's or Thomas Burnet's. Gosse's dilemma was to try to find a way to reconcile his science and his faith. He chose to challenge the rapidly growing support for evolutionists from the geological record.

In the years before the publication of *On the Origin of Species*, Charles Darwin had had to master the art of scientific politics. A hopeless champion for his own ideas, at least in public, Darwin's skill lay in behind-the-scenes proselytising, persuading his scientific colleagues (who were often opponents at first) and co-opting them to persuade others. How far or how quickly his ideas would have spread without others having become his public champions is hard to know. As it happened, it was disciples such as Huxley and the botanist J. D. Hooker who preached Darwin's new theory to the public.[136] Huxley had not always been a spell-binding orator although he was always persuasive. Unlike Darwin, who was born into ease if not luxury, Huxley had made his way in the world the hard way. He first trained as a doctor by becoming an apprentice in London's destitute East End. Lacking capital and influence, he signed on as a naval surgeon and used a voyage around the world in the surveying ship HMS *Rattlesnake* between 1846 and 1851 to collect specimens and develop ideas. As had Darwin fifteen years before, he returned a fully 'made' scholar and soon achieved a place as a scientist and a teacher.

Huxley had a favourite lecture – a 'Lay Sermon' – entitled *Essay on a Piece of Chalk*.[137] He would stand before an eager crowd and take a piece of common chalk from his pocket, asking the audience what it could possibly tell them about the history of the cosmos and of life on earth. The answer is that chalk (in those days, before blackboard chalk was an artificial, hypo-allergenic substance) represents the accumulation on an ancient sea bottom of the skeletons of countless billions of microscopic planktonic organisms that once inhabited vast tropical oceans that extended across the earth, from Europe and the Middle East to Australia and North America. Their skeletal remains consist of calcium carbonate and they are cemented together by chemical precipitations of the same substance. The chalk cliffs of the Kent and Sussex coast are old (about 135 million years): at the minutely slow rates of modern sedimentation, the beds that make them up – hundreds of feet thick – could only have been formed by accumulation over

Thomas Huxley

millions of years. And then were buried under hundreds of feet of other rocks, equally slowly built up.

Philip Gosse knew only too well what a piece of chalk looked like under a microscope and that the earth's crust consisted of thousands of feet of different rocks, some bearing fossils, others the remains of ancient lava flows, dust storms, water-borne sediments, and even ancient coral reefs just like those he had seen in Jamaica. All of these had been laid down over millions of years, compressed, folded, partially obliterated, sometimes redeposited in a new place, comprising a system for which the only possible analogy is the leaves of an ancient, mistreated book. How could Gosse explain away this all-too-solid evidence of the ancient history of the earth and its denizens? What did it have to say about the biblical account of creation in six days or the extent to which we could trust the 'document of revelation' concerning God's existence and wisdom? Surely science had shown, once and for all, that the earth was ancient and subject to law-driven changes. Whatever else it might, or might not, say about God, at least it said that the account in Genesis was not a literal account, but a fable – in fact, a rather nice metaphor.

Gosse's answer cost him dearly. The dilemma figuratively tore him – scientist and fundamentalist Christian – in half. In a classic example of *ad hoc* reasoning, he explained away all this appearance of change in a book entitled *Omphalos*,[138] the Greek for 'navel', and in that one word is contained the core of Gosse's argument. It is the old conundrum: did Adam have a navel? If God created Adam as the first man out of nothing, Adam would have had no need for a navel, since he had never been connected by an umbilical cord to a mother. Nor indeed had Eve, of whose origin Genesis gives two accounts. Nor indeed (remembering that the Bible tells us that God made man in his own image) would God physiologically have needed a navel.

Gosse simply asserted that at the moment of creation, just as God made Adam with a navel, he also made the earth with all its complex layers, its faults, every one of its fossils, volcanoes in mid-eruption and rivers in full spate carrying a load of sediment that had never been eroded from mountains that had never been uplifted. Similarly, at that instant, every tree that had never grown nevertheless had internal growth rings; every mammal already had partially worn teeth. He created rotting logs on the forest floor, the rain in mid-fall, the light from distant stars in mid-stream, the planets part-way around their orbits ... the whole universe up and running at the moment of creation: no further assembly required.

Such an argument, of course, can never be beaten. It says that God has created all the evidence that supports his existence and (shades of Hume) all the evidence that appears to cast doubt on it. Equally, of course, a theory that explains everything explains nothing. *Omphalos* is untestable and therefore one cannot concur rationally with its argument; you must simply close your eyes and believe. Or smile.

Over the years, Gosse's argument has been bowdlerised to the slightly unworthy proposition that God set out the geological record, with all its evidence of change, in order to test man's faith. It was, therefore, the ultimate celestial jest and cruel hoax. This was about as far from Gosse's pious intention as Darwin's impious theory. As for what Paley would have made of *Omphalos* – I like to think he would have rejected it, but kindly, for he was a kind man. Victorian England not only rejected it, they laughed at it cruelly. Gosse became overnight a broken man, his reputation as a scientist in shatters.

But nothing is as simple as it ought to be. A community that mocked *Omphalos* and had no problem in coming to terms with the even more difficult issue of cosmology, still could not come to terms with geology. In fact, whether in Paley's time or in Darwin's, or indeed our own, one of the oddities in the history of interplay between science and religion is that cosmology never seems to have become as serious a threat to revealed religion as natural science. When pressed, people often revert to believing two things at once. The evidence that the universe is huge and ancient can be assimilated seemingly without shaking the conviction that the earth itself is 6,000 years old and that all living creatures were created over a two-day period. For example: 'The school books of the present day, while they teach the child that the earth moves, yet assure him that it is a little less than six thousand years old, and that it was made in six days. On the other hand, geologists of all religious creeds are agreed that the earth has existed for an immense series of years.'[139] These last words were written in 1860 and appear in a work that arguably presented a greater threat to the Established Church than the evolutionism of

Erasmus Darwin, Lamarck, Robert Chambers or even Charles Darwin. *Essays and Reviews* was an example of the enemy within, a compilation of extremely liberal theological views by noted churchmen and academics. Among their targets was the unnecessary and outmoded belief in miracles and the biblical account of the days of creation. That battle is still being fought.

Perhaps it is a question of scale: millions of miles and even a single light year are virtually incomprehensible to us. Whenever cosmology threatens to become dangerously comprehensible it can be pushed off into the never-never land of concepts and incomprehensible hypotheses (like Einstein's relativity, which makes no difference at all to how long a kettle takes to boil). This was how those ideas took hold in the first place. There was no magical moment at which the discoveries of Copernicus and the observations of Kepler and Tycho Brahe suddenly took over Western thought (as if in some kind of paradigm shift). The Copernican heliocentric universe eased itself in, being accepted first on the basis of being a heuristic approach to calculating predictive star maps for navigation – a tool to use, rather than something to believe in and die for (at least after Giordano Bruno, who was burned at the stake in 1600 for heresy). Galileo first presented his telescope (a 'spy-glass') to the Doge of Venice as a military instrument rather than an aid to natural philosophy. After Galileo, the elements of the new cosmology became acceptable, even in Catholic countries, so that when Descartes and Newton extended things to the earth beneath our feet, the surrender was more or less complete – and still it could be conveniently relegated to the realms of speculation. But it turned out that the new ideas on evolution could not be consigned so conveniently to the category of 'interesting but unproven.'

CHAPTER ELEVEN

Good and Evil: Concerning the Mind of God

'I form the light and create the darkness: I make peace, and create evil: I the Lord do all these things.'

Isaiah 45:7

'No man hath seen God at any time.'

John 1:18

Charles Darwin's transformation from Christian believer to agnostic 'crept over [him] at a very slow rate'. For a long time he held on to a position as a theist, being prepared to believe in the existence of a God but rejecting both the Old Testament, 'from its manifestly false history of the world . . . and from its attributing to God the feelings of a vengeful tyrant', and also the miracles of the New Testament.

The more we know of the fixed laws of nature the more incredible do miracles become . . . the men at that time were ignorant and credulous to a degree almost incomprehensible to us . . . the Gospels cannot be proved to have been written simultaneously with the events . . . by such reflections as these, which I give not as having the least novelty or value, but as they influenced me, I gradually came to disbelieve in Christianity as divine revelation . . . But I was very unwilling to give up my belief . . . [A] source of conviction in the existence of God, connected with the reason

232

and not with the feelings, impresses me . . . [as does] the extreme difficulty, or rather impossibility of conceiving this immense and wonderful universe, including man with his capacity of looking far backwards and far into futurity, as the result of blind chance or necessity. When thus reflecting I feel compelled to look to a First Cause having an intelligent mind in some degree analogous to that of man; and I deserve to be called a Theist. This conclusion was strong in my kind about the time, as far as I can remember, when I wrote the *Origin of Species*; and it is since that time that it has very gradually with many fluctuations become weaker . . . and I for one must be content to remain an Agnostic.[140]

With the words 'a First Cause having an intelligent mind' Darwin, forty-eight years after having read Paley's *Natural Theology*, reflected the critical goal of that book and also the part that tells us most about the state of thinking about science and religion in Paley's time. Oddly, however, judging by most commentators, many readers of *Natural Theology* had given up long before they reached what Paley thought was the essential part of his argument, which comes in the twenty-third chapter. Certainly the vast majority of modern readers seem to read the watch analogy, find it impressive or nonsensical depending on their point of view, and then become bogged down in the welter of anatomical detail that Paley layers onto his argument. Either that, or they are so dismayed by the direction in which Paley finally directs his principal thesis that they pretend that he has merely contented himself with proofs of the 'existence of God'.

Paley's supporters often demand that we should not condemn him for over reaching himself by attempting to define the 'intelligence' and analyse the 'personality' of the deity.[141] However, to limit Paley to his watch analogy is rather like a loyal supporter of a soccer team asking us to admire the way in which a player has struck the ball, the power of the shot, and the graceful arc it describes, without worrying ourselves about where it was aimed, whether it scored the goal, swerved embarrassingly to left or right, fell short, or – we should allow this too – was blocked by the opposing team. On the contrary, simply establishing the existence of God was not enough for Paley. For him, there was much more work to be done, more proofs to be given, many enemies to be slain, more metaphysical swamps to wade through. As the subtitle of *Natural Theology* states, he proposed to discover for us the *Attributes of the Deity*. There could be no more fundamental subject for enquiry than to probe the very nature of God himself.

The philosopher of science Karl Popper famously reasoned that in science one cannot 'prove' anything. Science, to be worth anything, must be framed as hypotheses that can be falsified; it is not, then, a set of protocols for discovering facts but instead a process of raising propositions and trying to disprove them. Remembering Hume's objection to Newton's first rule of reasoning, (and see p. 19) we can only keep testing our propositions time and time again. If we always come up with the same answer, we are on safe ground in proceeding as if something were true. If not, we keep looking. Postmodern philosophers have seized on this and come up with the notion that there are no facts, only strongly (or weakly) supported opinions. In this they are doing no more than following Hume's scepticism. The problem for scientists is to judge when an opinion or a belief becomes so consistently, so unanimously, supported that it must be treated as fact. Yet, as all the principal characters in this book discovered at one time or another, the hardest thing to know is the extent of one's ignorance. Newton thought it was worthwhile experimenting with alchemy, and the case of Dr Plot's strange fossil thigh bone is even more instructive.

Good and Evil: Concerning the Mind of God

When the first more complete skeletons of dinosaurs were found in the early nineteenth century, having no prior experience of dinosaurs, scientists were forced to reconstruct them according to the models they did know (Newton's first law of reasoning). The very first dinosaur, megalosaurus, that so perplexed Dr Plot and so delighted Dr Buckland, happens to be one of the many huge extinct carnivores that walked on their hind legs. Nothing like that existed in nature, so Buckland and his contemporaries quite reasonably reconstructed it as a sort of overgrown quadrupedal hyaena. Full-sized replicas of megalosaurus and other dinosaurs, made in this mode for the Great Exhibition of 1851, still survive in the park at Crystal Palace in south London. They serve as a nice reminder of how far from the 'truth' science may be at any given time. And they certainly warn us that current opinion (perhaps particularly about dinosaurs) is similarly contingent, provisional, or even downright inadequate.

Even if it meant straying far beyond his proclaimed boundaries of safe empirical enquiry and objective knowledge, in order to elucidate the 'personality' of God and thus to impute characteristics to a God that 'no man hath seen, at any time', Paley had to develop his arguments for the existence of the designing, intelligent hand of God into a demonstration of what that intelligence really was. The ineffable had to be explained in terms of the material. But the trouble with using terms like 'intelligent', 'design', and 'good' as anything more than metaphors is that it becomes easy to think that they can be applied literally in a human sense. While to portray God as too mysterious is always to risk making him too remote, making him too accessible and too much like us risks trivialising him. For the mainstream Christian, the role of Jesus was in many respects to bridge this gap. Jesus was sent in human form with all its strengths and weaknesses, as exemplified in the ultimate sense by the temptations and his cruel death. The fact of the human

235

Jesus helps to justify an anthropomorphic concept of God. Our natural theologian predecessors then had no qualms (or at least they could settle their qualms) about using their logic to follow the tradition of nearly two thousand years and to ascribe to God a series of human attributes. With typical assurance, Paley therefore gives us the title of his key chapter, nothing less than 'On the Personality of the Deity'. No serious reader could settle for 'The Argument Cumulative' when proofs of the very nature of God were in store.

The Bible tells us (Genesis 1:24) that 'God created man in his own image' (although there may be a great deal more truth in the Sunday School howler 'man created God in his own image'). For Paley, God really was a *person*, not some ineffable, immaterial force:

> Contrivance, if established, appears to me to prove every thing which we wish to prove. Among other things it proves the personality of the Deity, as distinguished from what is sometimes called nature, sometimes called a principle: which terms, in the mouths of those who use them philosophically, seem to be intended, to admit and to express an efficacy, but to exclude and to deny a personal agent. Now that which can contrive, which can design, must be a person. These capacities constitute personality, for they imply consciousness, and thought. They require that which can perceive an end or purposes; as well as the power of providing means, and of directing them to their end. They require a centre in which perceptions unite, and from which volitions flow; which is mind. The acts of a mind prove the existence of a mind; and in whatever a mind resides is a person. The seat of intellect is a person.

Here is everything that Hume had raged against. It is argument by analogy taken to the ultimate degree. In fact, Paley almost seems to vacillate with respect to the term 'a person', hinting instead that God may be something like

> The great energies of nature . . . known only to us by their effects . . . gravitation . . . We have no authority to limit the properties of mind to any particular corporeal form, or to any particular circumscription of space. These properties exist, in created nature, under a great variety of sensible forms . . . This sphere may be enlarged as an indefinite extent; may comprehend the universe: and, being so imagined, may serve to furnish us with a good notion, as we are capable of forming, of the immensity of the divine nature, i.e. of a being, infinite as well in essence, as in power . . .

But then he finishes the sentence '. . . yet nevertheless a person.'

If God is a person, what kind of person? Without getting meshed in all the metaphysical issues implied in the question of God's person-ness, Paley tried to summarise what the evidence of nature had to contribute to the subject. His argument *for* God had all along been the goodness and purpose of creation. On the other hand, since time immemorial, the most obvious argument *against* God's existence and source of doubt about his unique and universal goodness has been the manifest *imperfection* of nature, and especially the misery of a great deal of existence that tells of nothing but a bleak purposelessness. As Darwin wrote: 'it revolts our understanding to suppose that [God's] benevolence is not unbounded, for what advantage can there be in the sufferings of millions of the lower animals throughout almost endless time?' To say anything about God himself, Paley had therefore to tackle one of the most intransigent problems in theology. He had to explain all the material phenomena of disease and death, and the seven deadly sins of

237

social behaviour, in the same terms of God's goodness as all the wondrous complexity of form and function in nature.

Nature is not always as rosy as Ray and Paley portrayed it. Instead, life is often unrelentingly cruel, earnest, red in tooth and claw. The old enemy David Hume had written:

> Look around this universe. What an immense profusion of beings, animated and organised, sensible and active! You admire this prodigious variety and fecundity. But inspect a little more narrowly these living existences, the only beings worth regarding. How hostile and destructive to each other! How contemptible or odious to the spectator! The whole presents nothing but the idea of a blind nature, impregnated by a great vivifying principle, and pouring forth from her lap, without discernment or parental care, her maimed and abortive children ... [The circumstance] whence arises the misery and ill of the universe, is the inaccurate workmanship of all the springs and principles of the great machine of the universe.[142]

The situation is no better and perhaps even worse for the human species than for the putatively lesser forms of life. Nature is not a cosy story of universal goodness, sweetness and light. We adopt the noble lion as a metaphor for strength and bravery, but there is little nobility in being the deer (or child) that is ripped apart by the lion and eaten while its viscera are still quivering in the dust. It is hard to see a divine utilitarian goodness in venomous snakes, stinging wasps, mosquitoes and poisonous plants, or in leprosy, malaria and cancer, or in the miseries of old age and the death of the very young. For humans, ugliness, disharmony, war, tyranny, famine, viciousness, greed, racism, inter-religious and intra-religious conflict seem to be at least as common a part of our condition as goodness, happiness, peace and beauty.

Good and Evil: Concerning the Mind of God

This is the stalemate debated in every pulpit, denied at the hospital bed, elided at every graveside – an acid eating away at the faith of young and old. A benign and loving God has somehow to be squared with all the slings and arrows of outrageous fortune that flesh is heir to. If God has not created all this misery and evil, and if they do not flow as some natural consequence of his creation, we would have to accept that it has some other cause. In that case, God would not be the only First Cause, but one of many possible causes. Given the premises on which it was based, natural theology could not avoid the challenge of finding an explanation of this paradox, to provide a new explanation of why good and evil are equally God's work. This was its Achilles heel, and in the attempt to produce a rational scientific explanation of misery, want and evil, a door was opened for Darwin.

The first defence of every religious argument, faithfully repeated by Paley, is that everything happens according to a plan of God's making, a plan we are not capable of knowing or understanding: 'God works in mysterious ways, his wonders to perform.' In this view we must *in all faith* accept God's will no matter how appalling the manifestations of that 'will' might be. This is Gosse's *Omphalos* applied to metaphysics, and equally inconclusive, asking us to conspire against our rational selves and accept that, when disease or wickedness assail us, God's shifting benevolence towards us is part of his greater and unknowable rules. For Charles Darwin a serious personal crisis of faith came when his beloved daughter Annie died at the age of ten.[143] It was one of the last straws in the metaphysical questioning that began in his student days at Edinburgh. In his *Autobiography* – 'I can indeed hardly see how anyone ought to wish Christianity to be true' – he reflected on his loss of faith in a passage that reveals how conscious he was of Paley's natural theology and how comprehensively he rejected it: 'The old

239

argument of design in nature, as given by Paley, which formerly seemed to me so conclusive, fails, now that the law of natural selection has been discovered … There seems to be no more design in the variability of organic beings and in the action of natural selection, than in the course which the wind blows. Everything in nature is the result of fixed laws.' And as for universal goodness and purposefulness of nature, 'passing over the endless beautiful adaptations which we everywhere meet with, it may be asked how can the generally beneficent arrangement of the world be accounted for? Some writers indeed are so much impressed with the amount of suffering in the world, that they doubt if we look to all sentient beings, whether there is more of misery or of happiness; – whether the world as a whole is a good or a bad one.'

Parallel to the assertion of 'God's will' being done is the defence that everything can be explained in terms of goodness after all. Paley assured his readers that: 'The common course of things is in favour of happiness … happiness is the rule: misery, the exception.' Where the unbeliever or sceptic sees a world of pain and suffering, the believer will see 'the extensiveness of God's bounty'. Unhappily, here the great logician succumbed to the seduction of utilitarian argument and produced perhaps the least convincing moment of the whole natural theology movement: '*Shoals of the fry of fish … are so happy that they know not what to do with themselves … their vivacity, their leaps out of the water, their frolics in it … all conduce to show their excess of spirits.*' Aristotle could have told him that they were jumping out of the water to escape from predators attacking from underneath. The early bird catches the worm and the late worm gets caught.

A similar defence lies in an inverse logic that Paley was not averse to using, for example: 'to the imbecility of age, quietness and repose become positive gratifications'. Or, as he had written earlier in his *Moral Philosophy*, one could argue that if God 'had wished our misery, he might have made sure of his purpose ... He might have made, for example, every thing we tasted bitter; every thing we saw loathsome; every thing we touched a sting; every smell a stench.' This all seems very unsophisticated and would scarcely be worth repeating except as a prelude to what follows.

Robert Boyle (among many others) had long since insisted that God's works were not intended for 'the welfare of such and such particular creatures ... the welfare and interest of man himself (as an animal) must give way to the care that Providence takes of a more general and important nature or condition.'[144] Paley was willing to grasp this intellectual nettle. He tried to find uses and purposes in the roles of the disembowelled deer and even nettles, and his answers were surprisingly modern. At the risk of seeming to patronise, we must note his explanation of the curse of any North American or North Eurasian picnic. At first it seems absurd to see a divine purpose in mosquitoes, gnats and blackflies, but 'immense tracks of forest in North America would be nearly lost to sensitive existence, if it were not for gnats'. Science was moving forward slowly in the field of what we now call ecology and this allowed Paley to see a divine provision in the extraordinary fecundity ('superfecundity') by which what would otherwise be barren forests were filled with teeming life.

Significantly, here Paley was reflecting a move to a more modern way of thinking. Crude and simplistic, it heralded a sea change in the way that scholars saw the world. In this new interconnected nature a loss in one direction is a gain in another and life itself is balanced with death: 'without death there could be no generation, no sexes, no parental relations, i.e. as things are constituted, no animal happiness.' There are echoes here of what Robert Boyle had written more than a hundred years before: 'tame and fearful birds . . . seem designed to be the food of hawks, kites and other rapacious ones.' Paley put a softer gloss on it – his predators only culled populations of their weak and sick.

Paley had now shifted the terms of the argument so that everything good and intelligent stemmed from the First Cause while all that appeared unpleasant or unplanned stemmed from a Second Cause. In the process, Paley's tortured dancing on the heads of all these metaphysical pins is a pre-shadowing of modern ecological thinking and a metaphysical extension of Hooke and Newton's explanation of planetary orbits in terms of opposing forces, or Woodward's theory of matter, or Hutton's geology – it is the living world as a dynamic system of force and counterforce, of checks and balances. Now we can see why God appears to have created some inconveniences; they exist as part of what Paley called the great 'shifting economy' of nature. In this view, life is neither all good for all creatures, nor totally arbitrary and cruel. Life exists as a shifting economy in which all the properties a sentimentalist finds distasteful can be seen actually to work for the divine purpose.

At this point, even if he does conclude that this economy is directed at supporting human existence, the archdeacon is flirting, all unbeknownst, with some of the elements that will eventually become part of Darwinian theory. He then goes on to demonstrate how neatly a brilliant teacher can turn the tables on his adversaries; he finally takes up the great enemy of 'chance' and uses it to his own advantage. Having dismissed, or at least derided, the notion of chance as a cause of fundamental processes, he finds uses for random occurrences after all – when they are operating within the safe realm of Second Causes. 'Why should there be, in the world, so much, as there is, of the appearance of *chance*?' His first answer is that chance is normal and insignificant: 'there must be chance in the midst of design:

by which we mean, that events which are not designed, necessarily arise from the pursuit of events which are designed. One
man travelling to York meets another man travelling to London.
Their meeting is by chance, is accidental.' His second answer
is that what appears to be chance is in fact the result of natural
laws: 'the appearance of chance will always bear a proportion
to the ignorance of the observer. The Cast of a die, as regularly
follows the laws of motion, as the going of a watch: yet . . . we
call the turning up of the number of the die chance . . .' Then,
some chance events have a necessary and good (or at least
salutary) role: 'for example, it seems to be expedient that the
period of human life should be uncertain . . . were deaths never
sudden, they who are in health, would be too confident of life.'

However, as we follow this argument along, a jarring socio-
political agenda abruptly shows through:

> There are strong intelligible reasons, why there should exist
> in human society great disparity of death and station . . . for
> instance, to answer the various demands of civil life, there ought
> to be amongst the members of every civil society a diversity
> of education, which can only belong to an original diversity of
> circumstances . . . [Can this disparity] be better disposed of than
> by chance? Parentage is that sort of chance . . . It may be that the
> fortunes . . . of the father devolve upon the son . . . but, with
> respect to the successor himself, it is the drawing of a ticket in a
> lottery . . . Even the acquirability of civil advantages, ought, per-
> haps, in a considerable degree to lie at the mercy of chance . . . for
> the poor, that is, they who seek their subsistence by constant
> manual labour, must still form the mass of the community.

Thus natural theology is revealed as supporting social stability
and, ultimately, preservation of the oligarchy: 'There must
always be the difference between the rich and the poor.' God's

shifting economy of nature and the vicissitudes of chance all come together not only to create and preserve unchanged the natural world, but to justify the social and political status quo. As the Bible puts it: 'In a great house there are not only vessels of gold and silver, but also of wood and earth' (II Timothy 2:21). Kings, bishops, labourers and shopkeepers (and archdeacons like Paley) all have, through the chances of their birth, an appointed place that they must accept cheerfully and obediently. Sure enough, the charming children's hymn 'All Things Bright and Beautiful' has the same sentiment in the sting in its tail. Edited out of some modern hymnals is the following verse:

> The rich man in his castle
> The poor man at his gates
> God made them, high and lowly
> And ordered their estates.

CHAPTER TWELVE

Paley, Malthus and Darwin

'And God blessed them, and God said unto them, Be fruitful, and multiply, and replenish the earth, and subdue it.'
Genesis 1:28

'It is desirable that there should be, in any rank of society,
as little as may be of that luxury, the object of which is to
contribute to the spurious gratifications of vanity; that
those who are least favoured with the gifts of fortune,
should be condemned to the smallest practicable portion
of compulsory labour; and that no man should be obliged
to devote his life to the servitude of the galley slave, and
the ignorance of the beast.'
William Godwin, *The Enquirer*, 1797

At first sight there seems no obvious link between the preceding quotation from an atheistical journalist and a discussion
of Archdeacon Paley's analysis of the personality of God or
Charles Darwin's theories about evolution. The words come
from a work entitled *The Enquirer: reflections on education,
manners, and literature* by William Godwin and were written
in 1797 at the height of a fever of utopian thinking fueled by
liberal French authors such as Jean-Jacques Rousseau and the
Marquis de Condorcet. Utopianism fitted very well with ideas
of liberté, égalité, fraternité – although less well with the Terror
and Napoleonic imperialism.[145] The connection arises because
one of Godwin and Rousseau's friends was Daniel Malthus, a

gentleman-scholar who had an estate at Albury in Surrey where he maintained a literary salon for liberal thinkers. Rousseau visited London several times and in 1769 he and David Hume visited the Malthus household together. The son of the house, Thomas Robert, was three years old at the time. In the way that sons have, he grew up taking a contrary position to most things his father stood for, among them his utopian dreams. The son had a careful, analytical brain and for him the down-to-earth precision of numbers and geometry meant more than abstractions and dreamings.

The young Malthus attended Jesus College, Cambridge, graduating in 1788 with the highest possible academic distinction in mathematics. He took holy orders the next year and – like so very many before him – began life as a scholarly parson, first taking the curacy at Albury, the family home. Among other subjects, he was passionate about discovering the causes of poverty and, therefore, the means of ending it. This led him to take a close interest in demography, the mathematics of populations. The consequences of this made him famous – his name a household word (although one of opprobrium more often than praise) – and a key member of the small number of scholars upon whose work Charles Darwin's ideas depended. He was also a major influence on Paley, and here we begin to see disparate worlds colliding and, out of the resulting intellectual tumult, a new theory arising.

A lapsed dissenting clergyman, a republican, novelist, social philosopher and journalist, William Godwin was also the husband of the feminist author Mary Wollstonecraft, father of Mary Shelley (author of *Frankenstein*) and thus the father-in-law of Percy Bysshe Shelley. Godwin's popular writings summarised a whole movement of progressive thought and owed a lot to Rousseau and to Condorcet's *Esquisse d'un tableau des progrès de l'espirit humain* (1794). In his *Political Justice* of

William Godwin

1793, Godwin expounded a utopian idea of the perfectibility of man and society, and of equality among men, which he saw as driven by the economics of population growth. Having proceeded upward from the savage, man's continuation towards perfection must follow as a law of nature. Population growth meant a growth in labour, which was only good, and could only lead to greater wealth and improvement for all – providing, of course, that institutions changed accordingly. Eventually there would be no crime, no war, no need for governments, nor even any sexual intercourse.[146]

The young curate Thomas Malthus was living at home in 1797 when Godwin published *The Enquirer*, another very popular treatment of this subject. Father and son argued bitterly over Godwin's criticisms of capitalism (particularly over the essay *On Avarice and Profusion*) and his dream of a world in which everyone lived at the level of the highest common denominator. Malthus Junior was quite sure that the opposite would be true – humanity would never rise significantly beyond the *least* common denominator. Indeed, everything in contemporary life pointed in that direction. The proliferating labouring classes of Europe lived in an abject poverty that made a lie of any forecast of a quick translation of population growth into utopian ease and plenty.

Demography and population had long been a subject of interest to natural philosophers. Earlier (in Chapter 6) we noted that Thomas Burnet chose a clever argument with which to open his *Sacred Theory of the Earth*. A newly created world was essential if the biblical view of creation was to be upheld and if Bishop Ussher's chronology was a correct reading from that record. Against Aristotle's view that the earth was eternal, old and unchanging, Burnet had to find features that had definitely changed during the course of the earth's known history. One of those was the size of the human population: ''Tis certain the World was not so populous one or two thousand years since, as it is now, seeing 'tis observ'd, in particular Nations, that within a space of two or three hundred years, notwithstanding all casualties, the number of Men doubles [if] the Earth had stood from Everlasting, it had been overstockt ere this ... Whereas we find the Earth is not yet sufficiently Inhabited, and there is still room for some Millions.'

Burnet's authority for the doubling-time of the human population was probably none other than Bishop Ussher's analysis of biblical genealogies and the growing contemporary interest in the demographics of populations. Population growth allowed Burnet to argue that the world was changing, but that it had not been changing for very long. For example, at the Norman Conquest the population of England and Wales was around one million; by the end of the seventeenth century it was around five million; a hundred years later it was seven million; in 1800 it was ten million (today it is some 60 million).

As is usual with Burnet, what seemed to him obvious was for others contentious. William Whiston almost reflexively opposed Burnet on the subject of human numbers. 'The inhabitants of the Earth were before the Flood vastly more numerous than the present Earth either actually does, or perhaps is capable to contain and supply.' He calculated the pre-deluge world population at 8,232 million and that of 1700 as 350 million. On another tack, John Ray's protégé William Derham, no doubt with his patron's detestation of Burnet in mind, saw population sizes and growth rates very much as a matter of divine wisdom and intervention:

> The whole Surface of our Globe can afford Room and Support only to such a number of all sorts of Creatures, And if by the doubling, trebling, or any other Multiplication of the Kind, they should encrease to double or treble that numbers, they must starve, or devour one another. The keeping therefore of the Balance even, is manifestly a Work of Divine Wisdom and Providence. To which end, the great Author of Life hath determined the Life of all Creatures to such a Length, and their Increase to such a Number, proportional to the Use in the World. The Life of some Creatures is long, and their increase but small, and by that means they do not overstock the world ... to balance the Stock of the Terraqeous Globe in all Ages, and Places, and among all Creatures, that is an actual Demonstration of our Saviour's Assertion ... even a sparrow ... doth not fall on the Ground without our Heavenly father.[147]

Here, then, we see the beginning of yet another demonstration that a part of the earth's internal economy was in balance. Derham even had an ingenious explanation of the fact that in England, observably in modern times, 'by special Providence, a few more still are born than die'. It is nothing less than a divine strategy allowing England to 'supply unheathful Places, where Death out-runs Life; to make up for the Ravages of great Plagues, and Diseases, and the Depredations of War and the Seas; and to afford a sufficient number for Colonies in the unpeopled Parts of the Earth'. He continued with reference to 'the Asiatick, and other more fertile Countries, where prodigious Multitudes are yearly swept away with great Plagues, and sometimes War, and yet those Countries are so far from being wasted, that they remain full of people'. Derham then went further and offered an explanation of the ancient puzzle and threat to the Bible's literal veracity: the apparently nonsensical ages for various Biblical characters. Where other authors had tried to find some calculus by which to adjust a biblical year to a modern one, Derham argued that the longevity of the early fathers was a special device for population growth: 'In the beginning of the World, and so after Noah's Flood, the Longevity of Men, as it was of absolute Necessity to the more speedy peopling of the new World, so is a special Instance of the Divine Providence.' Here is the solution to the puzzle of Methuselah's age (969 years) or even Abraham's. Longevity produced a deficit of deaths relative to births and so the population necessarily increased more quickly.

> And the same Providence appears in the following Ages, when the World was pretty well peopled, in reducing the common Age of man then to 120 years, [Genesis 6:3] in proportion to the Occasions of the World at that Time . . . And lastly, when the World was fully peopled after the Flood . . . and so down to our present time . . . the lessening of the age of Man to 70 or 80 Years . . . by this means, the peopled World is kept at a convenient Stay, neither too full or too empty.

(The logic of this fails to explain why, if population increase had been important in the beginning, Adam waited until he was 150 to have his first-born, or Methuselah until he was 187.) Needless to say, the argumentative Whiston had already staked his claim to an alternative theory on the ages of the ancients, once again with matters changing at the Flood. Before the deluge, the atmosphere was 'rare and thin' and conducive to long, healthy lives. Afterwards it changed to 'a Thick and Gross Consistency, from an equability . . . to extremity of Heat and Cold . . . to a mix'd and confus'd *Composition* or *Atmosphere*, wherein all sorts of *Effluvia*, Sulphureous, Nitrous, Mineral and Metallick &c. are contained'. Therefore we are now weaker and, as a result, more short-lived.

All that was the old style of reasoning, with strong utilitarian overtones. It fitted perfectly into the late-seventeenth-century world of sacred theories. Thomas Malthus, a modern thinker, not held back by the weight of convention, moved the whole subject from these kinds of treatments to a more scientific approach that depended on numbers rather than narratives. Consumed by a deeply Christian interest and applying his mathematical skills to the subject, he delved deep into the causes of poverty via the facts and principles of demography. In a typically careful and organised manner, he wrote out the simple elements of an argument against Godwin's utopian ideas for his father (one wonders if father and son argued by exchange of memoranda). As they debated, Malthus Senior urged his son to publish these notes, which he did anonymously in 1798 as *An Essay on the Principle of Population, as It Affects the Future Improvement of Society, with some Remarks on the Speculations of Mr Godwin, M. Condorcet, and other Writers.*[148] This pamphlet proved surprisingly popular – a measure of the extent of the poverty that he addressed. So Malthus produced a second edition (really a new book) in 1803, fleshing out his ideas with a mass of demographic data and anecdotes from his extensive travels.[149] In 1805 he married and started teaching history and political economy at the college of the East India Company at East Grinstead, Surrey, a post he held until his death.

Thomas Malthus

In his *Essay on the Principle of Population*, Malthus built on the work of earlier scholars who had made much out of analyses of births, deaths and numbers of offspring, read from parish registers. His great contribution to the discussion of population was to distil all these analyses into a two-part basic principle, applicable on a global scale. Firstly, empirical data showed that 'Population, when unchecked, increases in geometric ratio. Subsistence increases only in an arithmetic ratio. A slight acquaintance with numbers will show the immensity of the first power in comparison with the second ... That population cannot increase without the means of subsistence is a proposition so evident that it needs no illustration.'[150] But, since populations over most of the world are, on average, not increasing geometrically, 'the great question, then, which remains to be considered, is the manner in which this constant and necessary check upon population practically operates.'

Malthus's conclusion was that there must be 'a strong and

constantly operating check on population from the difficulty of subsistence. This difficulty must fall somewhere and must necessarily be severely felt by a large portion of mankind.' By the time he had written the second version of his *Essay*, he had worked out that there must be several kinds of checks acting to control population. The negative checks were 'all unwholesome occupations, severe labour, and exposure to the seasons, extreme poverty, bad nursing of children, great towns, excesses of all kinds, the whole train of common diseases and epidemics, wars, plague, and famine'. But positive checks also existed in the form of moral restraint and 'enlightened' human behaviour. Given the times and his own sensitivities, Malthus had to be a little delicate in listing these latter but they included delayed marriage, delaying the starting of a family, and various 'improper arts' and 'irregular connections' (abortion, contraception and prostitution) of which he thoroughly disapproved (sometimes Malthus is wrongly described as promoting contraception).

Once again we see a balance of forces, for growth in population and against it, from which Malthus's grim statistics led him to a single principal conclusion: that 'the superior power of population cannot be checked without producing misery or vice'. One does not have to read far into Malthus's short essay to discover how bleak was his opinion of the potential for improvement of society. 'The principal argument of this essay ... tends [to show] the improbability that the lower classes of people in any country should ever be sufficiently free from want and labour to obtain any high degree of intellectual improvement.' Where Godwin had dreamed of a world utopia, Malthus had articulated a natural, law-like, material obstacle to the perfectibility of man and society. The earthly paradise projected by Godwin – at the extreme of which people would eventually become immortal and passion between the sexes would become extinct – was impossible on all counts.

The bitter truth that Malthus preached was that the balance between increase and control acted to maintain populations and nations in a state of gross servitude, inequality and unhappiness. Nature's laws maintained a state of abject misery where 'the superior power of population [produced] misery or vice'. Man was destined to live like the animals, constantly to struggle to survive at a level just above subsistence. If this were not so, God would have set us on a different path from the beginning. Where David Hume had written: 'Every wise, just and mild government by rendering the conditions of its subjects easy and secure will always abound most in people, as well as in commodities and riches,'[151] Malthus argued precisely the opposite: the population principle made inevitable an hierarchically structured society with the lower strata grounded in misery and 'the necessity of a class of proprietors and a class of labourers.' Malthus, a decent and indeed a Christian man, believed that 'this must certainly be considered as an evil, and every institution that promotes it essentially bad and impolitic. But whether a government could with advantage to society actively interfere to repress inequality of fortunes, may be a matter of doubt.'

While doubtful about the capacity of governments to deal with demographic inevitabilities, Malthus was no apologist for the kind of capitalism that grinds the faces of the poor. He was preoccupied with finding the means of alleviating the poverty – first of a rural, and then an industrial underclass – by identifying and promoting the positive checks to population growth. His *Essay* stands within a long debate about poor laws that in England extends at least as far back as the reign of Queen Elizabeth I, who was among those who tried to alleviate poverty without institutionalising it, to help the needy without removing their incentive to work, to help indigent women without encouraging them to have larger families. After Elizabeth, various poor-law reforms came and went, usually with the result of exacerbating the situation. The subject attracted the attention of dreamers and realists, revolutionaries and reactionaries alike. For example, Condorcet had proposed something quite modern: 'A fund should be established which should assure to the old an assistance, produced, in part, by their former savings, and, in part, by the savings of individuals who in making the same sacrifice due before they reap the benefit of it.' (Malthus considered this 'absolutely nugatory'.) In Britain the first National Insurance Act was eventually passed by the Liberal government of Herbert Asquith in 1911.

As his detractors were only too happy to point out, Malthus's ideas owed a lot to his extensive reading of David Hume, Adam Smith, and reformers like Robert Wallace and Joseph Townsend (for example, the latter's *Dissertation on the Poor Laws by a Well-Wisher to Mankind*, 1786[152]). In dismissing Malthus, Karl Marx complained: 'If the reader reminds me of Malthus, whose "Essay on Population" appeared in 1798, I remind him that this work in its first form is nothing more than a schoolboy-ish, superficial plagiary of de Foe, Sir James Steuart, Townsend, Franklin, Wallace &c, and does not contain a single sentence thought out by himself.'[153] Marx might have missed the irony that his own ideas could be also found elsewhere, for example in none other than Godwin, who had argued in his *Political Justice*: 'There is no wealth in this world except this, the labour of man. What is misnamed wealth, is merely a power invested in certain individuals by the institutions of society, to compel others to labour for their benefit.'[154]

Just as Adam Smith had earlier inquired into 'the nature and causes of wealth', Malthus had set out to make a science of 'the nature and causes of poverty'. In the process he became 'the best abused man of his age' and 'Malthusian', like its sister term 'Darwinian', became an epithet for those exigencies that constantly challenge the best hope of the human spirit. At first acquaintance, William Paley found himself naturally opposed to Malthus's dismal logic, although he had no quarrel with the premise from which his conclusions flowed: namely that 'all animals, according to the laws by which they are produced, must have a capacity of increasing in a geometrical progression'. Apart from any moral reservations he might have had about Malthus's hard logic and harsh conclusions, for a natural theologian any property of nature must in principle lead to good; anything that involved human increase – 'go forth and multiply' – must be especially good. Filling the world with Englishmen (as Whiston had argued) must also be good. But Paley was won around, or perhaps turned around, because Malthus – 'dismal' or not – provided a scientific basis for the balance of apparent good and evil found in God's creation. Paley was more than happy to join Malthus in preaching a population theory, because Malthus seemed to have proved the inevitability of social inequality. He paraphrased Malthus freely:

The case is this. Mankind will in every country *breed up* to a certain point of distress. That point may be different in different countries or ages ... the order of generation proceeds by something like a geometrical progression. The increase of provision, under circumstances even the most advantageous, can only assume the form of an arithmetic series. Whence it follows, that the population will always ... pass beyond the line of plenty, and will continue to increase til, checked by the difficulty of producing subsistence. Such difficulty, therefore, along with its attendant circumstances, *must* be found in every old country; and these circumstances constitute what we call poverty, which necessarily imposes labour, servitude, restraint ... It seems impossible to people a country with inhabitants who shall all be in easy circumstances.

Once again we have been shown the last piece of the puzzle, that final element of Paley's not-so-hidden agenda. His world view, his notion of the nature and role of God, was not built simply upon Christian charity but also upon a concept of social and political stability, with the Church at its centre. Change – progress, freedom, self-improvement and primitive ideas about evolution – not only challenged men's view of God but also threatened the social-political fabric. If change was possible, either driven by chance or (perhaps especially) if it was driven wholly or in part through human volition, if the status quo was not naturally law-given, then men and women might aspire to and actually change their station, as in the despised philosophies of Rousseau and Tom Paine. Malthus's world may have been miserable for the many (and pleasant for the few) but it was above all a stable world. All the factors, the checks to population, that produced stability in numbers also tended to produce stability in social systems. Natural theology had thus been extended into a sophistry that was far removed from the imperatives of the watch analogy.

Was Malthus being cynical or realistic when he wrote: 'Man cannot live in the midst of plenty. All cannot share alike the bounties of nature. Were there no established administration of property, every man would be obliged to guard with force his little store. Selfishness would be triumphant'? The same would be true, of course, in the case of general want. The point Malthus was making was that the natural laws of population dictated that 'established administration' must not be overthrown.

Was Paley being realistic or opportunistic when he repeated Malthus's precepts?

> It is of the nature of property, not only to be irregularly distributed, but to run into large masses ... There must always there-

257

fore be the difference between the rich and the poor; and this difference will be the more grinding, when no pretension is allowed to be set up against it . . . So that the evils, if evils they must be called, which spring either from the necessary subordinations of civil life, or from distinctions which have . . . grown up in most societies, so long as they are unaccompanied by privileges injurious or oppressive to the rest of the community, are such as may . . . be endured, with very little prejudice to their comfort.

Or when he wrote:

The advantages of wealth . . . are not greater than they ought to be. Money is the sweetener of human toil; the substitute for coercion . . . With respect to station . . . distinctions of this sort are subject more of competition than of enjoyment . . . it is not by what the *lord-mayor* feels in his coach, but by what the apprentice *feels* who gazes on him, that the public is served . . . Command is anxiety; obedience ease . . . The poet asks, 'What is grandeur; what is power?' The philosopher answers, 'Constraint and plague'.

Reading this two hundred years later, one recoils. No wonder Marx abhorred natural theology, Paley and Malthus. Nonetheless, we must find it in our hearts to excuse some of the worst of natural theology's excesses because we must consider the times in which Paley wrote. It is easy for us to criticise, insulated as we are against the horrors of the rest of the world, even though (and perhaps in part because) newspapers, radio and television bring them to us daily. In 1800, Paley and his contemporaries faced the horrors of the Revolution in France and the immediate prospect of it overspilling into their homelands. They had just endured the trauma of a bitter war that pitted them against their American cousins. They lived in explosive years of war-time poverty and depression when the price of a loaf of bread was greater than the agricultural worker's daily wage. For Paley and Malthus, the constraints of poverty were only too keenly felt and the benefits of behavioural restraint and the scientific revolution on agricultural production were as yet unknown. But the revolutionary potential of any philosophy of change was only too real.

Natural theology (or at least Paley's version) found a valuable ally in Malthus. And this produced one of life's supreme ironies: there is a direct link between Paley's argument *for* the deity and Darwin's contribution to the arguments *against*. In adopting Malthus's ideas so strongly and so early, Paley helped promulgate them and ultimately contributed to the promotion of atheism in the form of the evolutionary theory of Darwin, who acknowledged that both Paley and Malthus had had a profound influence on his thinking. We know that Darwin had read Paley at Cambridge and in his *Autobiography* he admitted that initially he had been persuaded by the arguments of natural theology and '[the] argument of design in nature, as given by Paley, which formerly seemed to me so conclusive . . .'[155] Darwin also stated that in October 1838, 'I happened to read for amusement Malthus on *Population*, and being well prepared to appreciate the struggle for existence which everywhere goes on from long-continued observation of the habits of animals and plants, it at once struck me that under these circumstances favourable variations would tend to be preserved and unfavourable ones to be destroyed. The result of this would be the formation of new species. Here, then, I had at last got a theory by which to work.'[156]

Some authors have read this to mean that this was his first encounter with Malthus. But that is quite impossible. As the son of the Whig doctor Robert Darwin and grandson of Erasmus, and as the nephew (and future son-in-law) of a major figure in British industry (Josiah Wedgwood), Darwin could not possibly have avoided knowing at least the elements of Malthus's arguments. As the son of socially aware Unitarians and a nephew of the Wedgwoods, he could scarcely have been unaware of the social context. As a Whig student at Cambridge under the wing of the ultra-liberal Robert Stevens Henslow at the time the Reform Bill of 1831 was being debated (with the new Poor Law Bill to follow in 1834) it is unthinkable. But even if he had lived in some kind of cocoon, protected from any unwelcome intrusion of news about the condition of his fellow man, having read and 'almost memorised' Paley, he had seen all of Malthus there.[157]

As noted before, Darwin's theory of natural selection depends on the following chain of reasoning: living creatures always have the capacity to produce more offspring than can reach their own reproductive age; in populations where numbers are in balance a large proportion of all offspring must therefore die before reproducing. Each of the offspring of any sexually reproducing set of parents differs from the next. Overproduction plus a changing world create a series of filters that select which offspring are fittest to survive (the last part of the title of Darwin's book is *the Preservation of favoured Races in the Struggle for Existence*). Most of this is present in Paley, although none was original even with him. When Paley had justified 'the system of animal hostilities' – prey being eaten by a predator, venomous animals, and so on – he found in the natural order of things a great good. He challenged the reader: would he care to see 'a world filled with drooping, superannuated, half starved, helpless and unhealthy animals?' Paley did not leave the matter there. 'But to do justice to the question, the system of animal *destruction* ought always to be considered in strict connexion with another property of animal nature, viz. superfecundity. They are countervailing qualities. One subsists by the correction of the other.' And things were arranged by compensation, so that 'an elephant produces but one calf . . . a butterfly lays six hundred eggs . . . defenseless-ness and devastation are repaired by fecundity.' Here then is Darwin's overproduction neatly laid out (and also established as a necessary trait required to maintain a balance of population numbers).

From his reading of Malthus, Paley had anticipated two of Darwin's premises; overproduction (superfecundity) and the shifting economy of the struggle for existence. In so doing he established that adaptation is not just a state or condition, but also a process. Darwin took Malthus and Paley and turned their ideas upside down. Instead of simply contributing to a balance of forces maintaining the status quo, 'superfecundity' and the 'system of animal hostilities' (the struggle for survival), which they had thought precluded change, turned out to be its driving engine.

Darwin's theory, with its foundation in Malthus's strict mathematical laws, proved a body blow to natural theology. Despite the efforts of luminaries such as William Whewell[158] in mid-century, the popularity of natural theology declined after Darwin. Although many tried, it was not possible to enlist natural selection on the side of the angels by construing it as the result of God-given natural laws; Darwin's version of evolution, although he saw it as 'the result of fixed laws', depended too much on chance events, especially in the production of variation. In the years after 1859, the watershed date in evolution's respectability, adherents to natural theology increasingly came from more fundamentalist groups, such as the staunchly Calvinist palaeontologist Hugh Miller in Scotland, whose books reached a huge audience. The argument from design, in its elemental version of 'irreducible complexity', is today principally favoured by various groups of anti-evolutionists and modern Creationists, offshoots of fundamentalist Protestantism in the USA.[159]

When all is said and done, nothing in the decline of natural theology *precludes* a personal belief in the First Cause. As a last resort, there is a very simple rhetorical strategy: whatever science discovers, the latter-day natural theologian – in all honest faith – may simply add 'and don't forget that all this simply proves God's benevolence'. When science articulates a process, the believer may still insist that 'God made that possible'. Wherever science produces a beginning, religion can always answer 'and God was before that'. Here, for example, is a typical example of post-Darwinian natural theology, dating from 1930. *Science Rediscovers God* starts quite promisingly; the Reverend Ronald Campbell Macfie wrote that life had arisen in a volcanic context:

> All life today . . . depends on the manufacture of carbon and protein compounds in plant cells, and in the air around the volcano there would be plenty of the raw materials, carbon monoxide, carbon dioxide, methane, and there would also be a certain amount of nitrates dissolved in volcanic water or formed in the air by lightning and suitable bases for protein manufacture . . . my own theory [is] that the first living organism came to birth, or rather was created on the peak of a volcano – offspring of fire, and mire, and lightning.

That was the science, and it stood by itself and was testable. But he went on: 'But wherever it appeared there must have been some power at work not at work today, and a power exercised apparently by a prescient will.'[160] That was where the science ended and theology began. But it is not a true synthesis of science and theology. Neither proposition – that life began from simple molecules, and that God controlled the process – illuminates or validates the other.

All modern versions of the argument from design still use the wealth of complex detail in living systems as evidence for the necessity of an intelligent creator. And the origin of life itself has become the key to everything else. The argument from design is the foundation, for example, of Michael Behe's closely argued *Darwin's Black Box*.[161] The fine structure of a cell's cilium or a bacterial flagellum (the minute whip-like 'hair' by which they move) is revealed by the electron microscope to consist of a precise structural array of molecules that makes the workings of a watch look like a simple child's toy. As an example of something that could not have evolved through chance processes, Behe also cites the complex interactions of the biochemical pathways by which all living organisms work, paying special attention to mechanisms such as that by which blood clots as a defence against trauma. Modern molecular biology reveals a world of complex structures and mechanisms, acting at the microscopic level, that out-rivals the sort of examples from comparative anatomy and physiology that Ray and Paley used. Yet the conclusion from this is precisely that drawn by Paley: 'All of [these] parts are required to perform one function ... ciliary motion does not exist in the absence of microtubules, connectors, and motors ... Systems of horrendous, irreducible complexity inhabit the cell. The resulting realisation that life was designed by an intelligence is a shock to us.'[162]

The key word is 'irreducible'. In some modern versions of natural theology a great deal (perhaps most) of the properties of living systems may be allowed to have arisen through the action of Second Causes (and one may argue about the extent to which God supervises and directs the operation and outcomes of these natural processes), but there remains a core of complexity that is explained only by the operation of some external, designing Intelligence. This core is called 'irreducible complexity'. In fact, this key term can be used in two senses. The

263

first is that a system is irreducible if removal of a part prevents its function. Thus a football team is a complex social system, but removing one player will not necessarily prevent the team from playing. Similarly, if you remove the minute hand of a watch, the hour hand still gives you the time. However, if the balance wheel from the watch is missing, all is lost. This kind of axiomatic irreducibility is not controversial because it says nothing at all about how the complexity was caused: it is a way of defining one kind of complexity. There is, of course, also the opposite kind of complex system with a great deal of built-in redundancy – an army, for example, or almost any committee.

Something like a cilium or a hydra that may apparently be irreducibly complex in the functional sense, is quite different when seen from a developmental point of view. Any complex biological structure is assembled during reproduction and development from simple precursors, and those precursors from simpler ones: organs from tissues, tissues from cells, cells from molecules and atoms. This then introduces the second sense in which irreducibility is used. It derives from the philosophical concept of reduction, namely that any complex system can be reduced to the operation of simpler causes. Thus the parabolic trajectory of a projectile is the product of two straight-line forces acting on each other; physiological effects in the body are the result of processes acting at the molecular level; molecular processes are reducible to chemistry; chemistry to the underlying physics. For those who oppose the whole concept of evolution, 'irreducibility' in this sense is defined as a complexity that *cannot have evolved*. Ultimately it means that the first self-replicating molecule that began the whole history of life on earth could not have arisen by self-assembly from simpler molecules. Belief for the modern natural theologian can then be concentrated on the proposition that, if nowhere else in the entire world of nature, the origin of life required the hand of God.

Charles Darwin was well aware of the problem. He wrote in *On the Origin of Species*: 'If it could be demonstrated that any complex organ exists, which could not possibly have been formed by numerous, slight modifications, my theory would absolutely break down. But I can find no such case.' To most modern scientists, the hypothesis that there is a level of complexity that cannot be reduced to (or have arisen from) simpler component systems and subsidiary causes more closely resembles an assertion of faith than science. In the Behe quotation given earlier, the existence of a designing intelligence was presented as a 'realisation', not a logical conclusion from critical experiment. The whole concept depends on a negative premise (reducibility will not be discovered) that is not especially self-evident, given that the whole history of science has been to take us ever onwards in the discovery of causes and mechanisms, always unveiling deeper and deeper levels of structural and causal simplicity. Nonetheless, 'irreducible complexity' constitutes a fascinating recasting of the argument from design. The only way to counter it would be to show in the laboratory how life can arise from simpler non-living constituents without extra-natural (or divine) intervention. But that, even Hume would instruct us, would not constitute proof that life *did* arise that way.

So we have come full circle and are left with the safest foundation for theology, not in natural science, but in revelation and faith. The distinguished Harvard astronomer Professor Owen Gingerich, for example, is a strong modern supporter of natural theology. Revealingly, he writes: '*Having made the leap of faith* [emphasis added], [I] find the arguments from design very illuminating.'[163] In this sense, therefore, natural theology and natural religion – and particularly biblical literalism – have finished up very much like Gosse's *Omphalos*: untestable and therefore of no real value except to those who need no convincing.

CHAPTER THIRTEEN

The Beginning of the End

'[At Oxford this week] the chief cause of contention has
been the new theory of Development of Species by Natural
Selection – a theory open – like the Zoological Garden
(from a particular cage in which it draws so many laughable illustrations) – to a good deal of personal quizzing,
without, however, seriously crippling the usefulness of the
physiological investigations on which it rests.'

Athenaeum; 7 July 1860

'The mystery of the beginnings of all things is insoluble by
us; and I for one must be content to remain an Agnostic.'

Charles Darwin, *Autobiography*, 1879

When I was a student, one of the famous men in our department
published what he fondly hoped would be the definitive work
in his subject. An enormous tome, it covered matters from alpha
to omega, from the ancient Greeks to the present day. The
students closest to him had a little party with champagne to
celebrate the publication of 'the last word'. And so it seemed
for the next few years, except that the subject became less interesting; it seemed that all the questions had been answered. Then,
first slowly and soon quickly, the whole subject fell apart and
reinvented itself, with younger scholars and previously unthinkable ideas. The great man's work turned out simply to be a
summary of what we used to think. So it was for *Natural Theology* and natural theology. They could not secure an intellectual

position against all comers for all time. It took a while for Darwinism to be fully accepted but eventually, of the whole argument for natural theology, only the watch analogy and the argument from design, with all their limitations, remained. All else was swept away or eroded, drop by drop, fact by fact.

A great deal of the early storm of change washed at that rock of ages the University of Oxford, where William Buckland was an embodiment of natural theology, a leading scientist who was also a prominent cleric. For most of the first half of the nineteenth century, science at Oxford had been in a parlous state, despite the presence of Buckland and other luminaries such as Charles Daubeny. The consensus view was that 'the obstacle to the study of physical science at Oxford does not lie, as some people suppose, in any peculiar obstinacy or perverseness of the Oxford mind, but in the very nature of the place . . . Theology absorbs all other subjects.' Indeed, in the 1820s it was a sign of the dominance of Paley's *Natural Theology* that the displays of the old Ashmolean Museum (which at that time were very much dominated by natural history) had been arranged according to the sequence of topics in that book.[164]

In the 1840s, with science and science education becoming a national priority but Oxford still seriously lagging behind, a group of dons led by Daubeny began lobbying the university for the establishment of a new honours school of natural sciences and a new building. Such a building was in fact essential given the growth of Oxford's scientific collections, uncomfortably housed variously in the old Ashmolean building, the Clarendon building next door and Christ Church College. Housing for scientists was equally cramped and scattered, especially for the burgeoning subjects of chemistry and 'experimental philosophy' (physics). Remarkably, William Buckland, England's greatest palaeontologist, refused to support something so 'alien from what is thought to be the proper business of the University as natural history in any of its branches'. He thought the university in its formal teaching should instead stick to the classics and theology. It was not until Buckland had gone off to be Dean of Westminster that more modern views could prevail and science could be brought less stealthily to the curriculum. In 1849, Convocation (the parliament of dons) finally voted to allow the one great innovation that Buckland had resolutely blocked – a special building for the sciences. It was to be both a place of new classrooms, laboratories and offices and at the same time a great museum where the university's fabulous scientific collections (including Buckland's) would be available for teaching and study, and also open to the public.

Dr Henry Acland, Lee's Reader in Anatomy at Christ Church, led the effort for the new building and, as soon as the idea had been accepted, the question of the appropriate architectural style for the new museum became the next battleground: should it be built in the familiar classical Palladian style in which the new Ashmolean Museum had just been rehoused, or in the highly fashionable modern Gothic revival. The issue became the embodiment of what the architect Augustus Pugin had identified as the choice between decadent (classical) and modern (neo-Gothic).[165] One of the great protagonists for the neo-Gothic was the artist and critic John Ruskin, who happened to be an old friend and Christ Church member with Acland. On a narrow vote, the Rhenish-neo-Gothic design (by the Dublin firm of Dean and Woodward) won the day. Ruskin (who is often incorrectly credited as the architect) wrote to his friend Lady Trevelyan: 'Acland has his museum – the architect is a friend of mine – I can do anything I like with it.'[166]

Funds were found for the new building from the handsome balance in the University Press's Bible-publishing account. This was no paradox, because the building was intended by Acland and Ruskin to be a paradigm of natural theology, 'founded to bring the light and beauty and life of the works of God to their eyes.' After many trials and tribulations (including running out of cash, which left the building unfinished to this day), the University Museum opened in 1860 as one of the greatest glories of Victorian architecture – a poem, a secular cathedral in glass and iron, forming, with its perfectly Ruskian decorations in wrought iron and carved stone, a veritable textbook of nature, and the great material apotheosis of natural theology. However, the new displays were arranged along modern scientific lines, not by reference to Paley.

Samuel Wilberforce

A proud university invited the British Association for the Advancement of Science to hold its July meeting in the museum. The local host committee was chaired by the Bishop of Oxford, Samuel Wilberforce, a man of forceful personality and such rhetorical skill in turning an argument to any advantage that he was known universally as Soapy Sam. (The nickname also stemmed from his habit of wringing his hands when he spoke.)

Suffering perhaps from the disadvantage of being the son of a far more famous and accomplished personage, the abolitionist William Wilberforce, Sam was always on the lookout for personal as well as intellectual triumphs. But he was well qualified to bring the scientists to Oxford, having taken a First in mathematics at Oriel. And Wilberforce had a very powerful scientific ally in Richard Owen, the London palaeontologist and anatomist and deadly opponent of Darwin's ideas.

The year 1860 was to be momentous. Not only had Charles Darwin just published *On the Origin of Species*, in the spring of 1860 *Essays and Reviews* appeared. And while Darwin was propounding dangerous theories usurping God's role in creation, from within the Church had come new calls for abandoning claims for the literal truth of the Bible, especially Ussher's notion of the age of the earth. The religious liberals, many of whom were Oxford dons such as the Reverend Baden Powell, Savilian Professor of Geometry, were fed up with conducting a rearguard action against the obvious truths of science. Needless to say, Wilberforce was of the opposite persuasion. And all of this was set in the context of a schism within the Church of England between the 'anglicans' and the Oxford Movement, with its tendency towards Rome.

The scientists came to Oxford buzzing with talk of evolution and on Saturday 30 June, Wilberforce and Thomas Henry Huxley came face to face in a debate that has gone down in history as one of the turning points in evolutionary thinking. To put the session of Saturday 30 June into context, we have to start with Thursday 28th. In the zoology and botany section that day, Professor Daubeny read a paper *'On the final causes of the sexuality in plants, with particular reference to Mr Darwin's work "On the Origin of Species by Natural Selection"'*. In the ensuing discussion, with Huxley in the audience, Richard Owen made the mistake of countering with the absurdly exaggerated claim that the brain of a gorilla was more different from that of a human than from that of the lowest primate. As Huxley had been lecturing around the country in the previous two years, preaching the similarity of ape and human brains as evidence of evolution, this was a blatant challenge.[167]

First Huxley made the lofty and remarkably offensive observation that he 'did not think that a general audience, in which sentiment would unduly interfere with intellect, was a fit place for such a discussion'. Then he contradicted Owen flatly but politely. And, with that, increasingly partisan discussion of Darwin's new theory dominated the entire meeting. By Friday evening Huxley, exhausted by all the talk and argument, decided to decamp to meet his wife in nearby Reading. But he bumped into Robert Chambers on the street. Perhaps mindful of the way that Wilberforce had attacked Chambers at the 1847 meeting and certainly remembering how he had himself rather savaged *Vestiges of Creation* in print, Huxley reluctantly agreed to stay on for the final session.

On the Saturday morning, most of the great and near great of British science assembled in the museum's unfinished reading room. They included Dr Joseph Hooker, Darwin's long-time supporter and botanical muse, Admiral Robert FitzRoy, formerly captain of HMS *Beagle*, now chief 'statist' to the meteorological office of the Board of Trade, and John Lubbock, Darwin's friend and neighbour. Interestingly, Richard Owen did not attend. He had been scheduled to chair the session but instead asked the Reverend Robert Stevens Henslow, Darwin's old mentor from Cambridge, to take his place, perhaps shrewdly hoping to make the expected defeat of Darwinism the more complete. Wilberforce was there, of course, seated on the dais in his honorary official capacity. Only the eye of this particular hurricane, Darwin himself, was – typically – absent. With his intestines shredded by the pressures of public notoriety and

271

private uncertainty, he was busy taking a cure at Dr Lane's Hydropathic Clinic in Richmond.

Contrary to popular legend, the intended centrepiece of Saturday's session was neither Wilberforce nor Huxley, although the subject matter was very much Darwinian. The chemist Professor John W. Draper of New York University was allotted pride of place to read his paper, 'On the Intellectual Development of Europe, considered with reference to the views of Mr Darwin and others, that the progression of organisms is determined by law'. *Jackson's Oxford Journal* (7 July) reported: 'Great interest was created . . . by a paper read by Professor Draper.' In fact, he spoke for too long and was very dull and his contribution to the proceedings is always discounted in view of the fireworks that followed. But his paper (and his 1864 book, *A History of the Intellectual Development of Europe*[168]) shows how quickly Darwinian theory had encouraged the entry of an evolutionary metaphor of adaptation and environment into what we would call social and political science.

After Draper, Henslow called upon the Reverend Richard Cresswell, who spoke denying any parallel between the intellectual development of man and the physical development of lower animals. Sir Benjamin Brodie (President of the Royal Society) focused on human intelligence being 'identical to the Divine Intelligence . . . [and] the absurdity of supposing the source of Divine power dependent on the arrangement of matter'. Then a 'Mr Dingle' was shouted down by the students. At this Henslow stated that he would only allow the floor to those with arguments, not 'for mere declamation'.

Now Wilberforce, with assumed reluctance, accepted the invitation to speak,

[condemning] the Darwinian theory as unphilosophical; as not founded upon philosophical principles, but upon fancy, and he

272

denied that one instance had been produced by Mr Darwin on the alleged change from one species to another had ever taken place. He alluded to the weight of authority that had been brought to bear against it – men of eminence like Sir B. Brodie and Professor Owen being opposed to it, and concluded, amid much cheering, by denouncing it as degrading to man, and as a theory founded upon fancy, instead of upon facts.[169]

Wilberforce (well coached by Owen), had all his facts in place. His skilled use of rhetoric, now strictly logical, now witheringly dismissive, always flamboyant, carried the majority of the audience along, the ladies in the window waving their white handkerchiefs, the students in the rear cheering and jeering, the clerics smugly applauding. At the end of this all-out attack, Wilberforce added the one rhetorical flourish that has gone down in history, supposedly asking Huxley sarcastically whether 'it was through his grandfather or his grandmother that he claimed descent from a monkey'.

This time-honoured version of Wilberforce's jibe did not appear anywhere in print until 1900 (in Leonard Huxley's *Life* of his father[170]). The account given by an eyewitness in a contemporary letter is subtly different: 'Up rose Wilberforce and proceeded to act as the Smasher. The white chokers [clergymen] who were present cheered lustily . . . as Samuel rattled on – "He had been told that Professor Huxley had said that he didn't see that it mattered much to a man whether his grandfather were an ape or no! Let the learned Professor speak for himself" and the like.'[171] The *Athenaeum*'s report was in the same vein: 'The Bishop of Oxford came out strongly against a theory which holds it possible that man may be descended from an ape. But others – conspicuous among these, Prof. Huxley – have expressed their willingness to accept, for themselves, as well as for their friends and enemies, all actual truths, even the last

273

humiliating truth of a pedigree not registered in the Herald's College.'

These accounts have a greater ring of authenticity than the traditional version, created thirty years later. Both suggest that the ape-grandfather metaphor had arisen on previous occasions – if not in Thursday's discussion, then even earlier. Even if Wilberforce had been prompted by Owen, the use of 'grandfather' was most likely Huxley's own coinage, here turned back against him.[172] Indeed, to have queried the legitimacy of Huxley's grandfather would have been bad enough, but for Wilberforce to have mentioned Huxley's grandmother (if in fact he did) would have been a particularly crude and shocking insult in decorous Victorian times.

Whatever he said, Wilberforce sat down to tumultuous applause, the uproar perhaps giving Huxley a moment to plan his reply, most of which was only what he had been repeating all through the meetings. When he rose, the room fell to silence. First, 'in a calm and dispassionate and argumentative speech', he skilfully defended the structure of Darwin's argument. Then he produced his famous riposte:

Huxley – young, cool, quiet, sarcastic, scientific in fact and in treatment . . . gave his Lordship such a smashing . . . This was the exordium 'I asserted, and I repeat, that a man has no reason to be ashamed of having an ape for a grandfather. If there were an ancestor whom I should feel shame in recalling, it would rather be a *man*, a man of restless and versatile intellect, who, not content with an equivocal success in his own sphere of activity [Huxley later disclaimed the word 'equivocal'], plunges into scientific questions with which he had no real acquaintance, only to obscure them by an aimless rhetoric, and distract the attention of his hearers from the real point at issue by eloquent digressions and skilled appeals to religious prejudice'.

Another witness recalled: 'Prof. Huxley replied that he would rather have for his grandfather an honest ape low in the scale of being than a man of exalted intellect and high attainments who used his power to pervert the truth.'

This created an even greater sensation. Lady Brewster (wife of the philosopher Sir David Brewster), secured a minor place in history by fainting. Even so, in those days Huxley was no great orator and many of the audience never heard the end of his famous remark. For Joseph Hooker, at least, it was by no means the *coup de grâce*. He wrote to Darwin, rather uncharitably: 'Well, Sam Oxon got up and spouted for half an hour with inimitable spirit, ugliness and emptiness and unfairness . . . Huxley answered admirably and turned the tables, but he could not throw his voice over so large an assembly nor command the audience . . . he did not allude to Sam's weak points nor put the matter in a form or way that carried the audience.'

In the version that has become legend, the meeting ended with Huxley. In fact, it continued. Next Henslow recognised none other than Admiral FitzRoy, who 'regretted the publication of Mr Darwin's book, and denied Prof. Huxley's statement, that it was a logical arrangement of facts.' Professor Beale 'pointed out some of the difficulties with which the Darwinian theory had to deal' and Darwin's friend and neighbour Sir John Lubbock 'expressed his willingness to accept the Darwinian hypothesis in the absence of any better'. All this time (he claimed), Joseph Hooker had been growing quietly more furious. Two days later, he wrote to Darwin:

Now I saw my advantage; I swore to myself that I would smite that Amalekite, Sam, hip and thigh . . . and I handed my name up to the President as ready to throw down the gauntlet . . . then I smashed him amid rounds of applause. I hit him in the wind at the first shot in ten words taken from his own ugly mouth

275

Before Darwin

Joseph Hooker

. . . I then proceeded to demonstrate in as few more: (1) that he could never have read your book, and (2) that he was absolutely ignorant of the rudiments of Bot. Science. I said a few more on the subject of my own experience and conversion and wound up with a very few observations on the relative positions of the old and new hypotheses . . . Sam was shut up – had not one word to say in reply, and the meeting *was dissolved forthwith*.

One is struck not only by the unrestrained venom of Hooker's language – the phrase 'his own ugly mouth' has a distinctly twentieth-century ring – but also by the fact that he did not miss a chance to jab at Huxley: 'Huxley, who had borne all the previous brunt of the battle, and who never before (thank God) praised me to my face, told me it was splendid and that he did not know before what stuff I was made of.' Nor did Hooker mention the ape-grandfather exchange which was, after all, the point on which Huxley later claimed his victory. One senses a certain rivalry as to which disciple Darwin should love best.

In the immediate aftermath of the meeting, all sides claimed to have won the day. Wilberforce wrote: 'On Saturday Professor Henslow who presided over the Zoological Section called on me by name to address the Section on Darwin's theory. So I could not escape and had quite a long fight with Huxley. I think I thoroughly beat him.' Huxley declared himself 'the most popular man in Oxford for a full four & twenty hours afterwards'. And Hooker claimed: 'I have been congratulated and thanked by the blackest coats and whitest stocks in Oxford.'

The other curious aspect of the debate is that the biblical references are a bit suspect. When first insulted by Wilberforce, Huxley supposedly whispered to his neighbour, 'The Lord hath delivered him unto my hand.' However, Huxley only remembered this some thirty years after the event.[173] Hooker set out to 'smite the Amalekite, Sam, hip and thigh', but biblical smiting by hip and thigh was performed by Samson on the Philistines (*Judges* 15:8), not on the Amalekites. On the other hand, when in Judges 7:15 Gideon is told 'The Lord hath delivered unto your hand the host of Midian', the 'host' in question were the Amalekites and the Midianites (who, with 'all the children of the east, lay in the valley like grasshoppers for multitude'). Of course, Judges and the first and second books of Samuel are full of the smiting, slaying and casual dismemberment of Amalekites, Midianites and Philistines, carried out fast and furiously by Gideon, Samson, Saul, David, et alia, so the confusion may be understandable. Soapy Sam would not have got it wrong, of course. But it seems possible that Huxley's 1891 remembrance of the debate had been coloured by a reading of Hooker's letter to Darwin.

With all the references to Philistines and Amalekites, the irony of the Oxford Museum having been built with Bible funds was complete, while Buckland was presumably spinning in his grave (seven miles distant). But no matter who had won the so-called debate, and even if a lot of the myth is just that, a tide had turned. Science everywhere progressed, even at Oxford (where Buckland's successor, John Philips, remained a life-long opponent of Darwinism), and it became harder and harder to keep a foot in both the scientific and religious camps.

Ruskin's later life was troubled and unhappy. Even the museum building, which he had always thought more than any other encapsulated his own aesthetic and cultural theories, came to repel him. Modern science, as conducted in the museum's laboratories, and particularly the theory of evolution and the vivisectionist practices of experimental physiology, disgusted him, as did railways, reductionism, Descartes and Protestantism. The building that had been built to glorify God's works became anathema to him. It was an embodiment of natural selection: '[a] narrow space . . . given to contemplation, [of] the Devil's working . . . through disease, and his triumph over [the works of God] in death'.

The doorway to the Oxford University Museum is still carved with Adam and Eve ('pretty Eve', Ruskin once wrote, 'always such a good bas relief, but then there is never anybody to match'). A tangled tree of life leads upwards from Adam and Eve on either side of a central angel. The angel holds the Book of Nature in his right hand while the left holds a strange disc. On close examination, this disc can be seen to represent the field of view of a microscope and the image is that of a dividing cell. The great temple to natural theology had been embellished with an image of the scientific mystery of generation, a reference (perhaps inadvertent) to the scepticism of Hume and emblematic of the new science, of which evolution and Charles Darwin's natural selection became its major theoretical basis, while the

Oxford debate became one of its most enduring public-relations successes.

Almost every culture has a version of the ancient joke in which a lost tourist asks for directions from a grizzled local (whether in deepest Maine, Yorkshire, the Australian outback, or the Masai Mara). The local replies to the effect that 'yes, I know where you want to go, but you can't get there from here'. Natural theology was the last great attempt to find a comprehensive answer to the question 'Does God exist and what is his nature?' through the objective, empirical methods of science rather than through revelation, biblical exegesis or the inspiration of God's vicars on earth. Since Paley, the most effective role of science in this baffling intellectual landscape has been to redefine, actually to isolate, the questions to a central core for which it is unable to provide the answers. Or, perhaps more fairly, religion and science have not been able to agree upon the reformulation of a set of questions that they can attack jointly. Oxford's team of scientists and architects, with Ruskin egging them on, had dearly wanted to create a complete and perfect entity, a total union of science and theology – natural theology expressed in stone, glass and iron. In the end, art imitated life. As a work of art and architecture, the building is incomplete – some windows beautifully carved as Ruskin wished, others left plain when the money for the sculptors ran out. As a statement of natural theology and the perfection of God's creation, it therefore fails. But in its unfinished state it makes a different, more modern, metaphor about the work of science and the reconciliation of science with religion.

The Account of Creation in Genesis

As fewer and fewer people actually read the Bible any more, it may be useful to quote directly the first book of the Old Testament, against which early sacred theorists had to measure their ideas. (Some repetitions may be omitted without detracting from the glory of the King James Bible.)

1 In the beginning God created the heaven and earth.
2 And this earth was without form, and void; and darkness was upon the face of the deep. And the Spirit of God moved upon the face of the waters.
3 And God said, Let there be light; and there was light.
4 And God saw the light, that it was good; and God divided the light from the darkness.
5 And God called the light Day and the darkness he called Night. And the evening and the morning were the first day.
6 And God said, Let there be a firmament in the midst of the waters, and let it divide the waters from the waters.
7 And God made the firmament, and divided the waters which were under the firmament from the waters which were above the firmament: and it was so.
8 And God called the firmament Heaven. And the evening and the morning were the second day.
9 And God said, Let the waters under the heaven be gathered together into one place, and let the dry land appear: and it was so.
10 And God called the dry land Earth; and the gathering together of the waters he called Seas; and God saw that it was good.

11   And God said, Let the earth bring forth grass, the herb
     yielding seed, and the fruit tree yielding fruit after his own
     kind, whose seed is in itself, upon the earth: and it was so.
13   And the evening and the morning were the third day.
14   And God said, Let there be lights in the firmament of the
     heaven to divide the day from the night, and let them be for
     signs, and for seasons, and for days, and years:
16   And God made two great lights; the greater light to rule the
     day, and the lesser light to rule the night: he made the stars
     also.
19   And the evening and the morning were the fourth day.
20   And God said, Let the waters bring forth abundantly the
     moving creature that hath life, and the fowl that may fly above
     the earth in the great firmament of heaven.
21   And God created great whales, and every living creature that
     moveth, which the waters brought forth abundantly, after their
     kind, and every winged fowl after his kind.
23   And the evening and morning were the fifth day.
25   And God made the beast of the earth after his kind, cattle
     after their kind, and every thing that creepeth upon the face of
     the earth: and God saw that it was good.
26   And God said, Let us make man in our image, after our
     likeness: and let them have dominion over the fish of the sea,
     and over the fowl of the air, and over the cattle, and over the
     earth, and over every creeping thing that creepeth upon the
     earth.
27   So God created man in his own image, in the image of God
     created he them; male and female created he them.
28   And God blessed them, and God said unto them, Be fruitful,
     and multiply, and replenish the earth, and subdue it . . .
31   And God saw every thing that he had made, and behold, it
     was very good. And the evening and the morning were the
     sixth day.

ACKNOWLEDGEMENTS

Two great books form the starting point for any student of this subject: John C. Greene's *The Death of Adam* (University of Iowa Press, 1959) and John Hedley Brooke's *Science and Religion: Some Historical Perspectives* (Cambridge University Press, 1991); my debt to these works is immeasurable. I am also grateful to John Hedley Brooke and Jim Kennedy (University of Oxford), Frank Turner (Yale University) and Eric Sperling (Stanford University) for reading the manuscript and saving me from numerous errors of commission and omission. Kristin Andrews-Speed, Bethia Thomas, Philip Powell, George McGavin, Kevin Walsh, Tony Fiorillo, Arthur McGregor, Vincent Strudwick, Dinah Birch, David Grylls, Eliza Howlett and Rennison Hall variously contributed essential discussions, information and assistance. Candace Guite and Mara Katvars at Christ's College, Cambridge, showed me the last Cambridge examination paper dealing with Paley's theology.

Geoffrey Thomas, President of Kellogg College, Oxford, provided a haven in which to write during a sabbatical leave. I thank the Bodleian and Radcliffe library staff and Stella Brecknell of the University Museum Library for all those little courtesies that make the difference between a good library service and a great one; as J. N. Brewer wrote in his *Oxfordshire* (1813), 'I have great pleasure in observing that at the Bodleian Library I found a constant wish to advance the object of my pursuit.' Jim Kennedy gave permission for me to photograph objects from the Geological Collections of the Oxford University Museum. Jim Bennett generously allowed me to photograph the lever watch by Josiah Emery (1784) from the collections of the University of Oxford Museum of the History of Science. I thank Mike Searle for permission to reproduce his photograph of the Himalayas at Chandigar and Mark Robinson for a photograph of the 'Temple' at Serapis. My wife, Linda Price Thomson, not only provided her usual moral support but created the portrait drawings. Felicity Bryan, literary agent

283

par excellence, was as ever a constant source of support and encouragement and Arabella Pike gave invaluable editorial advice. Portions of the final chapter were previously published in the journal *American Scientist* (copyright the author).

BIBLIOGRAPHY

Anonymous, 'Observations on the Nature and Importance of Geology', *Edinburgh New Philosophical Journal*, 1, pp. 293–302 (1826).

Arthuthnot, John, *An Examination of Dr. Woodward's Account of the Deluge*, 1697.

Baldwin, Stuart, *John Ray, Essex Naturalist: A Summary of His Life, Work and Scientific Significance*, Baldwin's Books, Essex, 1985.

Behe, Michael J., *Darwin's Black Box*, Simon and Schuster, New York, 1996.

Bentley, Richard, *Eight Sermons preach'd at the Honorable Robert Boyle's Lecture, in the first year, 1692*, London.

——, *The Folly and Unreasonabless of Atheism, in Eight Sermons*, London, 1724.

Boyle, Robert, *Free Enquiry Into the Vulgarly Perciev'd Notion of Nature*, London, 1685.

——, *Dissertation about the Final Causes of Natural Things*, London 1688.

Brooke, John Hedley, *Science and Religion: Some Historical Perspectives*, Cambridge University Press, 1991.

Brookes, Richard, *Natural History of Waters, Earths, Stones, Fossils and Minerals*, London, 1763.

Browne, Janet, *Charles Darwin: Voyaging*, Knopf, New York 1995.

Buchan, James, *Capital of the Mind*, John Murray, London, 2003.

Burbridge, David, 'William Paley confronts Erasmus Darwin: Natural Theology and Evolutionism in the Eighteenth Century', *Science and Christian Belief*, 10, pp. 49–71 (1998).

Burnet, Thomas, *The Sacred Theory of the Earth*, 1681, 1684, trans. and ed. Basil Willey, Centaur Press, 1965.

285

Campbell McFie, Ronald, *Science Rediscovers God: or, The Theodicy of Science*, Clarke, Edinburgh, 1930.

Cicero, Marcus Tullius, *De Natura Deorum*, or *On the Nature of the Gods*, 77 BC, ed. and trans. H. Rackham, Loeb Classical Library, New York, 1933.

Chambers, Robert, *Vestiges of the Natural History of Creation*, Churchill, London, 1844.

Clarke, Samuel, *A discourse on the Being and Attributes of God, being sixteen sermons preach'd in the years 1704 and 1705*, London, 1705.

Cudworth, Ralph, *The True Intellectual System of the Universe*, London, 1678.

Cutler, Alan, *The Seashell on the Mountaintop*, Dutton, New York, 2003.

Darwin, Charles, 'Autobiography' MS, Nora Barlow (ed.), Collins, London, 1958.

Darwin, Erasmus, *Zoonomia; or, the Laws of Organic Life*, London, 1794.

——, *The Temple of Nature*, 1803. London.

Dean, Dennis R., *James Hutton and the History of Geology*, Cornell University Press, 1992.

de Lamarck, Jean Baptiste, *Philosophie Zoologique*, Dentu, Paris, 1809.

Delsemme, Armand H., 'An Argument for the Cometary Origin of the Biosphere', *American Scientist*, 89, pp. 443–53 (2001).

Derham, William, *Physico-Theology*, London, 1713.

Descartes, Rene, *The Philosophical Works of Descartes*, Elizabeth Haldane and G. H. T. Ross (ed), 2 vols, Cambridge University Press, London, 1967.

Desmond, Adrian, *Huxley: From Devil's Disciple to Evolution's High Priest*, Addison Wesley, Reading, Massachusetts, 1997.

——, and James Moore, *Darwin: The Life of a Tortured Evolutionist*, Norton, New York, 1991.

d'Holback, Baron Paul H. T., *Systeme de la Nature*, 1770 (*The System of Nature or, the Laws of the Moral and Physical World*, 1844).

Bibliography

Drake, Ellen Tan, *Restless Genius: Robert Hooke and His Earthly Thoughts*, Oxford University Press, New York, 1966.

Drake, Stillman, *Galileo*, Oxford University Press, 1980.

Draper, John, *A History of the Intellectual Development of Europe*, Bell, London, 1864.

Edmonds, J. M., and H. P. Powell, 'Beringer's "Lugensteine" at Oxford', *Proceedings of the Geologists' Association*, 85, pp. 549–54. (1974).

Edwards, John, 'Brief Remarks upon Mr Whiston's New Theory of the Earth' in *The Existence and Providence of God*, London, 1697.

FitzRoy, Robert, 'A Very Few Remarks with Reference to the Deluge' in Robert FitzRoy and Philip Parker King, *Narrative of the surveying voyages of His Majesty's Ships Adventure and Beagle, between 1826 and 1836, describing their explorations of the southern shores of South America*, Henry Colburn, London, 1839.

Fortey, Richard, *Trilobites: Eyewitness to Evolution*, HarperCollins, London, 2000.

Franklin, Benjamin, 'Conjectures Concerning the Formation of the Earth', *American Philosophical Society Transactions* 3 (1793), pp. 1–5.

Fyfe, Aileen, 'The Reception of William Paley's *Natural Theology* in the University of Cambridge', *British Journal for the History of Science*, 30, pp. 321–35 (1997).

Galilei, Galileo, '*Il Saggiatore*' (1623), in ed. Drake, *The Controversy on the Comets of 1618*, University of Pennsylvania Press, 1960.

——, *Dialogo sopra i due massimi istemi del mondo*, or *Dialogue concerning the two major world systems*, 1632.

Gerard, John, *Historie of Plants*, London 1597; enlarged edition by Thomas Johnston, 1633.

Gingerich, Owen, 'Is There a Role for Natural Theology Today?' in

Murray Rae, Hilary Regan and John Stenhouse (eds), *Science and Theology: Questions at the Interface*, Clarke, Edinburgh, 1994.

Godwin, C. W., 'On the Mosaic Cosmogony' in *Essays and Reviews* (ed. anonymous), Longman Green, London, 1860.

Godwin, William, *An Enquiry concerning Political Justice, and its influence on general virtue and happiness*, Dublin, 1793.

——, *The Enquirer: Reflections on Education, Manners, and Literature*, Dublin, 1797.

Gosse, Edmund, *Father and Son: A Study of Two Temperaments*, Heinemann, London, 1913.

Gosse, Philip Henry, *Omphalos*, van Voorst, London, 1857.

Greene, John C., *The Death of Adam*, Iowa State University Press, 1959.

Grylls, David, *Guardians and Angels*, Faber and Faber, London, 1978.

Halley, Edmund, 'An Account of the Circulation of the Watry Vapours of the Sea, and of the Cause of Springs, Presented to the Royal Society', *Philosophical Transactions of the Royal Society of London*, 16, pp. 468–73 (1691).

——, 'Wise Men Now Think Otherwise: John Ray, Natural Theology and the Meanings of Anthropocentrism', *Notes and Records of the Royal Society of London*, 54, pp. 199–200 (2000).

Herschel, John F. W., *Preliminary Discourse on the Study of Natural Philosophy*, Longman, London, 1831.

Hooke, Robert, *Micrographia: or some physiological descriptions of Minute Bodies made by Magnifying Glasses, with Observations and Inquiries thereupon*, The Royal Society of London, London, 1665.

——, 'Lectures and Discourses of Earthquakes and Subterraneous Eruption. Explicating the Causes of the Rugged and Uneven Face of the Earth: and what reasons may be given for the frequent finding of shells and other Sea and Land Petrified Substances, scattered over the Terrestrial Superficies', in Richard Walker (ed.), *The Posthumous Work of Robert Hooke MD, FRS*, The Royal Society of London, 1709.

Bibliography

Hume, David, *Treatise of Human Nature* (1739), A. D. Lindsay (ed.), Everyman, London, 1956.

——, *Philosophical Essays Concerning Human Understanding* (1777), Erik Steinberg (ed.), Indianapolis, 1993.

——, *Dialogues Concerning Natural Religion* (1779), Martin Bell (ed.), Penguin, 1990.

Hutton, James, 'Abstract of a Dissertation, read in the Royal Society of Edinburgh upon the 7th of March and Fourth of April, 1785', *Royal Society of Edinburgh*, (1785).

——, 'The Theory of the Earth', *Transactions of the Royal Society of Edinburgh*, 1 (1788), pp. 209–304.

——, *An Investigation of the Principles of Knowledge and the Progress of Reason from Sense to Science and Philosophy*, Edinburgh, 1794.

——, *Theory of the Earth*, Edinburgh, 1795.

Huxley, Leonard, *Life and Letters of Thomas Henry Huxley*, London, 1900.

Huxley, Thomas Henry, 'Essay on a Piece of Chalk' in *Lay Sermons*, London, 1870.

——, *Lectures: The Principles of Biology*, Royal Institution, 1958.

Inwood, Stephen, *The Man Who Knew Too Much: The Strange and Inventive Life of Robert Hooke 1635–1703*, Macmillan, London, 2002.

Ito, Yushi, 'Hooke's cyclic theory of the earth in the context of seventeenth-century England', *British Journal for the History of Science*, 21, pp. 295–314 (1988).

Jahn, Melvin E., and Daniel J. Woolf, *The Lying Stones of Dr Johan Bartholomew Adam Beringer*, University of California Press, 1963.

Jardine, Lisa, *Ingenious Pursuits: Building the Scientific Revolution*, Little Brown, London, 1999.

Jefferson, Thomas, *American Philosophical Society Transactions*, 4, pp. 246–60. (1797).

Jones, Steven, *Almost Like a Whale: The Origin of Species Updated*, Doubleday, London, 1999.

Keynes, Randal, *Annie's Box*, Fourth Estate, London, 2001.

Lawrence, William, *Lectures of Physiology, Zoology, and the Natural History of Man*, Craddock, London, 1819.

Leigh, Charles, *The Natural History of Lancashire*, 1700.

LeMahieu, Dan L., *The Mind of William Paley: A Philosopher and His Age*, University of Nebraska Press, 1976.

Lhwyd, Edward, *Lithophylacii Botannici Ichnographia*, Oxford, 1699.

Lister, Martin, *Philosophical Transactions of the Royal Society*, 6 (1671).

Lyell, Charles, *Principles of Geology* (3 vols), John Murray, London, 1831–33.

McGregor, Arthur, and Abigail Headon, 'Reinventing the Ashmolean: Natural History and Natural Theology at Oxford in the 1820s to 1850s', *Archives of Natural History*, 27 (2000), pp. 369–406.

McMahon, Susan, 'John Ray (1627–1705) and the Act of Uniformity', *Notes of Records of the Royal Society of London*, 54, pp. 153–78 (2000).

Malthus, Thomas Robert, *An Essay on the Principle of Population, with remarks on the speculations of Mr. Godwin, M. Condorcet, and others*, Johnson, London, 1798.

——, *An Essay on the Principle of Population: or, a view of its past and present effects in human happiness; with an enquiry into our prospects respecting the future removal or mitigation of the evils which it occasions*, Johnson, London, 1803.

Manier, Edward, *The Young Darwin and His Cultural Circle*, Dordecht, 1998.

Marx, Karl, *Capital: A Critical Analysis of Capitalist Production*, London, 1867.

Matthew, Patrick, *Naval Timber and Arboriculture*, Harris, Edinburgh, 1831.

Bibliography

Neve, Michael, and Sharon Messenger, *Charles Darwin: Autobiographies*, Penguin, 2002.

Newton, Isaac, *Principia Mathematica* (1687), Cohen, I. Bernard, and Anne Whitman, California University Press, 2000.

Nieuwentyt, Bernard, *The Religious Philosopher, or the Right Use of Contemplating the Works of the Great*, 1709 (English translation 1719)

Numbers, Ronald L., *The Creationists*, University of California Press, Berkeley, 1993.

O'Dwyer, Frederick, *The Architecture of Woodward and Dean*, University of Cork Press, 1997.

Offray de la Mettrie, Julien, *L'homme machine*, 1748, ed. Paul-Laurent Assoum, Daniel-Gouthier, Paris, 1981.

Paine, Thomas, 'Of the religion of Deism compared with the Christian religion', in Daniel E. Wheeler (ed.), *The Life and Writings of Thomas Paine*, Parke, New York, 1908.

Paley, William, *The Principles of Moral and Political Philosophy*, London, 1768.

——, *A View of the Evidences of Christianity*, London, 1794.

——, *Natural Theology or Evidences of the Existence and Attributes of the Deity collected from the Appearance of Nature*, Hallowell, London, 1802.

Pearson, John, *Exposition of the Creed*, London, 1659.

Plot, Robert, *The Natural History of Oxford-shire*, Oxford, 1677, 1705.

——, *The Natural History of Stafford-shire*, Oxford, 1686.

Priestley, Joseph, review of Hume's *Dialogue Concerning Natural Religion* in *Letters to a Philosophical Unbeliever*, Johnson, London, 1787.

Pugin, Augustus, *Contrasts: or, A Parallel Between the Architecture of the 15th and 19th Centuries*, London, 1837.

Ray, John, *The Wisdom of God Manifested in the Works of His Creation*, Samuel Smith, London, 1691.

Rohl, David, *The Lost Testament: From Eden to Exile. A Five Hundred Year History of the People of the Bible*, Century, New York, 2002.

Romer, Alfred Sherwood, *The Vertebrate Body*, Saunders, Philadelphia, 1955.

Rudwick, Martin, *The Great Devonian Controversy: The Shaping of Scientific Knowledge Among Gentlemanly Specialists*, University of Chicago Press, 1985.

Ruse, Michael, *The Evolution Wars*, Rutgers University Press, New Jersey, 2000.

Ryan, William and Walter Pitman, *Noah's Flood: The New Scientific Discoveries about the Event that Changed History*, Simon and Schuster, New York, 1998.

Shapin, Steven, and Simon Schaffer, *Leviathan and the Air Pump*, Princeton University Press, 1985.

Smith, Jonathan, 'Philip Gosse and the Varieties of Natural Theology' in Linda Woodhead (ed.), *Reinventing Christianity*, Ashgate, 2001.

'Steno' (Niels Stenson), *De Solido Intra Solidum Naturaliter Contento Dissertationis Prodromus*, Florence, 1669. 'Prodromus' means preliminary treatise: unfortunately there was no follow-up. It was translated into English by Hans Oldenbergh in 1671 and a precis published in that year's *Philosophical Transactions of the Royal Society*, pp. 2186–90.

Stephen, Leslie, *History of English Thought* (2 vols), Smith Elder, London, 1876.

Thomson, Keith S., 'Vestiges of James Hutton', *American Scientist*, 89 (2001), pp. 212–214.

Thwaite, Ann, *Glimpses of the Wonderful: The Life of Philip Henry Gosse*, Faber and Faber, London, 2002.

Toland, John, *Christianity Not Mysterious*, London, 1696.

——, *Letters to Serena*, London, 1704.

Townsend, Joseph, *A Discourse on the Poor Laws by a well-wisher to mankind*, London, 1786.

Trembley, Abraham, *Memoires pour servir à l'histoire d'un genre de polypes d'eau douce, à bras en forme de cornes*, Leiden, 1744.

Bibliography

Uglow, Jenny, *The Lunar Men: The Friends Who Made the Future*, Faber and Faber, London, 2002.

Ussher, Bishop James, *Annales Veteris Testamentia prima mundi origini deducti*, or *Annals of the Old Testament deduced from the First Origin of the World*, 1650.

Vernon Jensen, J., 'Return to the Wilberforce–Huxley Debate', *British Journal for the History of Science*, 21 (1988), pp. 161–79.

von Humboldt, Alexander, and A. Bonpland, *Personal Narrative of Travels to the Equinoctial regions of America, during the years 1799–1804*, London, 1814–1829.

Whewell, William, *The Philosophy of the Inductive Sciences, founded upon their History*, London, 1840.

Whiston, William, *The genuine works of Flavius Josephus*, London, 1687.

——, *A New Theory of the Earth from its Original, to the Consummation of all things. Wherein the Creation of the World in Six Days, the Universal Deluge and the General Conflagration, as laid down in the Holy Scripture, are shewn to be perfectly agreeable to Reason and Philosophy*, London, 1697.

White, Gilbert, *The Natural History and Antiquities of Selborne in the County of Southampton*, White, London, 1788.

Wilde, Simon A., John W. Valley, William H. Peck and Colin M. Graham, 'Evidence from Detrital Zircons for the Existence of Continental Crust and Oceans on the Earth 4.4 Byr Ago', *Nature*, 409 (2001), pp. 175–91.

Woodward, John, *An Essay towards a Natural History of the Earth: as also of the Seas, Rivers, and Springs. Wilth an Account of the universal Deluge: and of the effects it had upon the Earth*, London, 1685.

Young, Davis A., 'The Biblical Flood as a Geological Agent: A Review of Theories' in *The Evolution-Creation Controversy II: Perspectives on Science, Religion, and Geological Eduction*, ed. Walter L. Manger, in *Papers of the Paleontological Society*, 5 (1999).

293

NOTES

1. Charles Darwin, 'Autobiography'. See also Neve and Messenger.
2. Pearson. *Exposition of the Creed*. London.
3. Herschel. *Preliminary Discourse*.
4. von Humboldt and Bonpland. *Personal Narrative*.
5. Paley, *Natural Theology*. All quotes from the 1826 edition, Hallowell, London (that being the edition most likely read by Charles Darwin as a student).
6. Charles Darwin, 'Autobiography'.
7. Fyfe, 1997.
8. Deuteronomy 6:16.
9. Luke 11:29.
10. LeMahieu. *The Mind of William Paley*.
11. Paley, *The Principles of Moral and Political Philosophy*.
12. Paley, *A View of the Evidences of Christianity*.
13. The timelessness of Paley's case may be indicated by the fact that two hundred years later, the Bishop of Saint Albans' Easter message for 2001, printed in the London *Sunday Times* under the subtle headline 'Yes, Jesus did rise from the dead, and I can prove it', begins by faithfully (so to speak) rehearsing Paley's *Evidences*.
14. Charles Darwin, 'Autobiography'.
15. LeMahieu.
16. Hume, *Philosophical Essays*.
17. Hume, *Treatise of Human Nature*.
18. Brooke, 2000.
19. There is the opposite play on words in Shakespeare's 'Now is the winter of our discontent/Made glorious summer by this sun of York.' (*Richard III*, I, I).
20. Drake. *Galileo*.

21. Newton. *Principia Mathematica*.
22. Galilei. *Il Saggiatore*.
23. Inwood. *The Man Who Knew Too Much*.
24. Shapin and Schaffer. *Leviathan and the Air Pump*.
25. Inwood.
26. Drake. *Restless Genius*. Jardine. *Ingenius Pursuits*.
27. Uglow. *The Lunar Men*.
28. Erasmus Darwin, *The Temple of Nature* (note this was actually published after Paley's *Natural Theology*).
29. Hutton, *Theory of the Earth*.
30. Boyle, *Dissertation*.
31. Hume, *Dialogues Concerning Natural Religion*.
32. Paine. *Of the religion of Deism*.
33. Toland, *Christianity Not Mysterious*.
34. Cudworth, *The true intellectual System of the Universe*.
35. Charles Darwin, 'Autobiography', where he cites these memories as part of the faith he lost.
36. Anonymous, *Quarterly Review*, 1802.
37. Stephen, *History of English Thought*. Vol. 2.
38. Behe. *Darwin's Black Box*.
39. d'Holback. *Systeme de la Nature*.
40. Nieuwentyt. *The Religious Philosopher*.
41. Cicero. *De natura Deorum*.
42. McMahon.
43. Gerard. *Historie of Plants*.
44. Baldwin.
45. Ray. *The Wisdom of God*.
46. In all quotations given in the book I have tried to keep the original spelling and punctuation as there is a particular pleasure in seeing things as closely as possible to the way they were written.
47. Boyle. *Free Enquiry*.
48. Some he was not averse to borrowing from others. In his *Dissertation about the Final Causes of Natural Things*, Robert Boyle had discussed the human skeleton: 'for the bones alone are reckoned to amount to three hundred . . . every one must

have its determinate size, figure, quality, consistence, situation, connexion, &c. and *any or all* of them together, must conspire to such and such determinate functions and use.' The same example appears almost verbatim three years later in Ray's *Wisdom of God*.

49. Some later editors of Paley's work, in an excess of zeal, advised their readers that Paley really meant 'all instances' (see the American Library Edition, edited by Elisha Bartlett, 1839).

50. Inwood.

51. Hooke, *Micrographia*.

52. Boyle, *Final Causes*.

53. Boyle left money in his will to endow an annual series of lectures on the subject of natural theology that were given between 1692 and 1732.

54. Clarke. *The Being and Attributes of God*.

55. Burbridge, 1998.

56. This has been described by Leigh van Valen as the Red Queen Effect. In *Through the Looking Glass*, by Lewis Carroll, the Red Queen said: 'It takes all the running you can do to keep in the same place.'

57. Jones.

58. Romer, *The Vertebrate Body*.

59. Bentley, 1724 'The Folly and Unreasonableness of Atheism' in *Eight Sermons*.

60. John Ray thought that erect posture was a function of our large and heavy heads. As for the fact that human children first move around on all fours, he explained that in terms of the unequal length of the limbs.

61. Behe.

62. Offray de la Mettrie, *L'homme machine*.

63. Lawrence, *Lectures on Physiology*.

64. Desmond and Moore, *Darwin*.

65. Trembley.

66. Hooke, 'Lectures and Discourses'.

67. Plot, *The Natural History of Oxford-Shire*, (all quotations

given here are from the second edition, 1705) and *The Natural History of Stafford-Shire*.

68. White, *Selborne*.
69. The case of Anne Green, executed for the murder of her child, appears to be the inspiration for a central storyline in Iain Pears' fascinating novel *An Instance of the Fingerpost* (1997, Jonathan Cape). This book gives a wonderful sense of the intellectual atmosphere of Plot's Oxford in the late sixteen hundreds.
70. Jahn and Woolf, *The Lying Stones*. Edmonds and Powell.
71. A flood occurs in several independent mythologies; see, for example, Ovid's *Metamorphoses* and the *Epic of Gilgamesh*.
72. Lister, pp. 2282.
73. I suppose it is yet another irony that the Portland stone walls of the modern Ashmolean Museum (built 1844) contain the imprints of thousands of Jurassic oyster shells – perfect 'impressions made directly in sediments'.
74. Fortey.
75. Lhwyd, *Lithophylacii Botannici Ichnographia*.
76. A further indignity was that in 1763 Richard Brookes, in his *Natural History of Waters, Earths, Stones, Fossils and Minerals*, turned the image of Plot's bone upside down and called it 'Scrotum humanum'. This notoriety no doubt contributed to its disappearance.
77. Jefferson.
78. One can perform an interesting mind experiment here: if Western science (the province of the fashionably derided dead white male) had evolved in Australia and China, say, rather than western Europe, would the story we now tell be different or the same? Would we see the same patterns but in different rocks, the same evolution but in different fossils, and above all, would we see the same age? The answer has to be yes. (See Rudwick, *The Great Devonian Controversy*).
79. Browne.
80. Lyell, *Principles of Geology*.
81. Burnet, *The Sacred Theory of the Earth*.

82. Everyone knows something of the story of Archimedes
    (*c.* 287–212 BC). Hiero, King of Syracuse, had asked him to
    discover whether a crown given to him was made of pure gold
    or was adulterated with silver. In his bath one day, Archimedes
    had the insight that the volume of water displaced by a
    submerged body is a measure of the weight of the body.
    Therefore, if one put the crown into water and measured how
    much water was displaced, one could compare that volume
    with the displacement caused by an equal weight of pure gold.

    A unit weight of any substance will displace a particular
    volume of water (or other given liquid). This measure is the
    relative density or specific gravity of the substance. Scientists,
    for all their vaunted objectivity, love to latch onto topics and
    work them to death. Newtonian gravity became a metaphor for
    dozens of unrelated subjects and, as we shall shortly notice,
    comets were the hot topic in the late 1600s. For those looking
    for a natural, lawful, ordering principal inherent in the
    properties of matter, specific gravity was a perfect tool. One
    could even assign a numerical value to this particular property
    and therefore order the different elements.

83. Just as there are two different accounts of the creation of
    woman, there are two different versions of the Flood in
    Genesis.

84. Greene, *The Death of Adam.*

85. Woodward, *Essay.*

86. Among those who quickly contested Woodward were John
    Arthuthnot (1697, *An Examination of Dr Woodward's
    Account of the Deluge*) and Charles Leigh (1700, *The Natural
    History of Lancashire*).

87. Young, '*The Biblical Flood as a Geological Agent*'.

88. The term geologist only came into use a hundred years later,
    but is convenient to use here.

89. Hooke, *Lectures and Discourses of Earthquakes.* Here Hooke
    applied the principle of a balance of forces directly to the
    matter of water, referring explicitly to his cosmological model:
    'Water is rais'd up in Vapours into the Air by one Quality and

precipitated down in drops by an other, the Rivers run into the
Sea, and the Sea replenishes them. In the circular Motion of all
the Planets, there is a direct Motion which makes them
indeavour to recede from the Sun or Center, and a magnetick
or attractive Power that keeps them from receding. Generation
creates and death destroys . . . All things always circulate and
have their vicissitudes.'

90. Halley, 1691.
91. Whiston, *The genuine works of Flavius Josephus.*
92. Whiston, *A New Theory of the Earth.*
93. Edwards.
94. There is now corroboration from modern science: see
    Delsemme.
95. Steno, *Prodromus.* pp. 2186–90.
96. Cutler.
97. FitzRoy.
98. *Ibid.*
99. Ito.
100. *Ibid.*
101. Toland, *Letters to Serena.*
102. Buchan, *Capital of the Mind.*
103. Hutton, 'Abstract of a Dissertation'.
104. Dean, *James Hutton and the History of Geology.*
105. Hutton, 1795. *Theory of the Earth.*
106. Hutton, 1788. '*The Theory of the Earth*' 209–304.
107. 1788. *Ibid.*
108. Hutton, 1795. *Theory of the Earth*, op. cit.
109. 1795. *Ibid.*
110. Hutton, 1788. *Theory of the Earth*, op. cit.
111. Recently, evidence of the very oldest rocks on earth, a vestige
     perhaps of creation, was discovered in Australia. See Wilde et
     al., and Thomson.
112. Franklin, 1793.
113. William Whiston, incidentally, had been sure that a close
     reading of the Bible showed that the ark had not landed on
     Mount Ararat but in the Caucasus.

Notes

114. FitzRoy, 1839.
115. Ryan and Pitman, *Noah's Flood* and see also *National Geographic*, December 2000.
116. Rohl, *The Lost Testament*.
117. Hooke, 'Lectures and Discourses'.
118. Anonymous, 'Observations on the Nature and Importance of Geology'.
119. de Lamarck, *Philosophie Zoologique*.
120. Toland, *Letters to Serena*.
121. Cicero.
122. Bentley. 1692 *Eight Sermons*.
123. Priestley.
124. Erasmus Darwin, *Zoonomia*.
125. Ruse, *The Evolution Wars*.
126. Jones.
127. Two notes need to be added here. Fitness can be measured by the relative contribution of a lineage to succeeding generations, but reproductive success is only a measure of fitness, not the sole contributor to fitness – otherwise, the world would be swamped with the weedy, prolific organisms. Secondly, it is sometimes argued that the expression of natural selection in the phrase 'survival of the fittest' is a meaningless tautology: the fit survive, the survivors are the fit. This is not a tautology but an axiomatic statement like 'the fastest runner wins the race'.
128. It is necessary also to allow for the operation of some random effects, for example a new species may arise or change be accelerated when very small numbers of a population become isolated, say on an island in the Galapagos chain, and become highly inbred.
129. Matthew, *Naval Timber*.
130. Hutton, *An Investigation Principles of Knowledge*.
131. Thwaite, *Glimpses of the Wonderful*.
132. Chambers, *Vestiges of the Natural History of Creation*.
133. Grylls, *Guardians and Angels*.
134. Thomas Henry Huxley, quoted in Desmond.

135. Edmund Gosse. *Father and Son*. For a more charitable view, see Thwaite and Smith, 2001.
136. Desmond. *Huxley*.
137. Thomas Henry Huxley, 'Essay on a Piece of Chalk'.
138. Philip Henry Gosse. *Omphalos*.
139. C. W. Godwin, 1860.
140. Charles Darwin, 'Autobiography'.
141. Behe.
142. Hume, *Dialogues*.
143. Keynes. *Annie's Box*.
144. Boyle, *Free Enquiry*.
145. William Godwin, *The Enquirer*.
146. William Godwin, *An Enquiry concerning Political Justice*. (In *The Enquirer*, op. cit., Godwin was more realistic about sex, seeing the commerce of the sexes as necessary.)
147. Derham, *Physico Theology*.
148. Malthus, 1798. *An Essay on the Principle of Population*.
149. Malthus, 1803. Prospects respecting the future removal or mitigation of the evils which it occasions.
150. It is important to re-emphasise that Malthus did not report that populations *actually* double every twenty-five years. Rather that was their *capacity* – as demonstrated in a few exceptional circumstances like the early European population of North America, where such a rate was actually recorded. Malthus's point was just the opposite – that populations very rarely did grow at such rates; instead they remained in a stable equilibrium.
151. Hume, *Political Discourse*.
152. Townsend, *A Discourse on the Poor Laws*.
153. Marx, *Capital*.
154. William Godwin, *An Enquiry concerning Political Justice*.
155. Charles Darwin, 'Autobiography'.
156. *Ibid*.
157. In the same way, we cannot be sure how much of Hume he read. But if he read *Natural Theology*, he was exposed to some at least of Hume's thinking (and see Manier).

Notes

158. Whewell, *The Philosophy of the Inductive Sciences.*
159. Numbers, *The Creationists.*
160. Campbell McFie, *Science Rediscovers God.*
161. Behe.
162. Behe.
163. Gingerich, 1994.
164. McGregor and Headon, 2000.
165. Pugin, *Contrasts.*
166. O'Dwyer, *The Architecture of Woodward and Dean.*
167. Thomas Henry Huxley, *Lectures.*
168. Draper, *A History of the Intellectual Development of Europe.*
169. *Jackson's Oxford Journal*, 7 July 1860.
170. Leonard Huxley, *Life and Letters of Thomas Henry Huxley.*
171. Contemporary sources for accounts of the debate are as follows: letters: Joseph Hooker to Charles Darwin, 2 July; John Richard Green to Sir William Boyd Dawkins, 3 July; Samuel Wilberforce to Sir Charles Anderson, 3 July; Balfour Stewart to David Forbes, 4 July and T. H. Huxley to Henry Dyster, 9 September; and *The Athenaeum*, 14 July; all 1860.
172. J. Vernon Jensen. 1988.
173. Thomas Henry Huxley, in Leonard Huxley.

INDEX

Acland, Henry 268
Act of Uniformity 65–6
Adam's navel 229
adaptation 63, 76–7, 78, 80,
    90–1
Agassiz, Louis 193
air, properties 34
Alps 146
amoeba 81
Arian heresy 89, 158
Aristotle 23–4, 39, 70, 101,
    144
Ashmole, Elias 122
Ashmolean Museum 122, 267
astronomy 24–5
asymmetry 88
atheism 200
atomism 14, 55
atoms
    Boyle 75
    Buffon 206
    Cudworth 56
    Democritos 30, 41
    Descartes 31, 41, 143
    eye 80
    life 99
    materialist philosophers 205
    Steno 166
    theory of change 103
    theory of matter 55

Toland 199
Whiston 161
'Autobiography' (Darwin) 4,
    239, 259

Bacon, Francis 31
balance of forces 28
Banks, Joseph 63
Barwick, Peter 153
Beagle, HMS 174, 175, 218
Beale, Prof 275
Behe, Michael 61, 263
Bentley, Richard 94–5, 205
Beringer, Johannes 125
Berkeley, Bishop 45
Bible 8, 9, 141, 250
    see also Genesis
biblical chronologies 116
bird migration 85–6
Black Sea 195
Bonnet, Charles 205–6
Boyle, Robert
    career 32–3
    Final Cause 42
    Hooke 34, 35
    nature 121
    philosophy 75–6, 241
Boyle Lectures 142, 205
Boyle's law 34

Brahe, Tycho 24
British Association for the
    Advancement of Science
    269, 270–6
Brodie, Sir Benjamin 272
Browne, William 105
Buckland, Mary Morland 192
Buckland, William 123, 134,
    191–3, 235, 267–8
Buffon, Compte du, Georges
    Louis leClerc 178, 206–7
Burnet, Thomas 141
    Telluria Theoria Sacra 62,
    142–51, 248

Cartesian *see* Descartes
causality 206
cause 101–2, 206
Cesalpinus 205
chalk 227
Chambers, Robert 216, 224,
    271
chance 71–2, 242–3
change
    earth 144
    evolution 201–2, 215
    Hume and Darwin 92
    natural philosophy 44
    Ray 172
    reproduction 212
    rocks 109
    theories 103
Christ's College, Cambridge 1,
    15, 20
Cicero 62, 200

Clark, William 135
Clarke, Samuel 89–90, 102,
    158, 217
classifications 70–3, 122–4,
    203
cleric-naturalists 7
comets 159, 161–2
compensation 89
complexity, irreducible 263–5
Concepcion, earthquake 175
continental drift 190
Copernican heliocentric
    universe 231
Copernicus, Nicholas 24
cosmology 43, 231
creation 115–16, 229
Cresswell, Richard 272
Cudworth, Ralph 55–7, 103,
    200, 205
Culpepper, Nicolas 67
cyclic theories 178, 179, 180,
    181

Darwin, Charles
    adaptation 63
    'Autobiography' 4, 239, 259
    Brazilian forest 60
    Cambridge 3, 4
    discoveries, personal
        consequences 20
    early life 1–5
    eye 93
    faith 2–3, 232–3, 239–40
    geology 138–9, 140
    irreducible complexity 265

Index

Lamarckism 215
Malthus 259
natural selection 4, 217, 218, 222, 224, 260–1
On the Origin of Species 270
Paley 5–6, 60, 204, 259
Parallel Roads of Glen Roy 194
reading 4–5
scientific politics 226–7
South America 174, 175
species, new 216
student 198
Darwin, Erasmus 4, 36–8, 103, 199, 210–12, 215
Daubeny, Charles 267, 270
deism 51
Democritos 30
Derham, William 173, 249
Descartes, René
atoms 41
Boyle 75
Cudworth 56
mind and soul 105
philosophy 28–31, 45
Principles of Philosophy 143
design argument *see* natural theology
development theory 7, 197
see also evolution
Dialogues Concerning Natural Religion (Hume) 49–50, 59–60, 90
dinosaurs 134, 235
diversification 217
diversity, life 203

DNA 100
Draper, John W. 272

earth
age 184
ancient 39, 114, 163, 228
Burnet 143–50
crust 189–90
Descartes 143
history 178, 186
Hutton 180–1
inner heat 156, 158, 163, 168, 187, 189, 190
structure 136
surface 109, 113
earthquakes 175, 179, 188, 190
Edwards, John 160
empiricism 31–2
Enlightenment
Bible 9
cleric-naturalists 7
deist scientists 31
France 54
free thinking 47
God, role of 55, 74
Paley 17
philosophy 12
Scotland 176
Epicureanism 28, 30, 41, 56, 75
Epicurus 147
epistemology 30
erosion 155, 163–4, 178, 181
Essay on the Principle of Population, An (Malthus) 251–4

*Essay towards a Natural
History of the Earth*
(Woodward) 153–8
evolution
  change 201–2
  Darwin, Erasmus 37–8, 199,
    210–12
  Lamarck 213
  natural selection 221
  pre-Darwin 55, 197–9, 201,
    224
  premises 217
evolutionary metaphor 272
experimentation 84
extinction 118, 135
eye 93–4
  human 80

faith 8, 46
Final Cause 42–3, 107–8
First Cause 40, 262
see also God
FitzRoy, Robert 140, 174, 175,
    191, 271, 275
Flood 148–9, 154–5, 161, 171,
    190–1, 195
forces, balance of 28
fossil bone 132–4
fossils
  Beringer 125
  Darwin, Erasmus 37, 210
  debates 132
  facts 118–19
  faunas 191
  Greeks and Romans 126

Hooke 113, 126–7
hunting for 119
Leonardo da Vinci 126
Lhwyd 130–1
Plot 122–4, 128
questions raised by 117–18
Ray 130
record 202–3
Smith 136
see also palaeontology
Fox, William Darwin 2
France 54
Franklin, Benjamin 189
free thinking 47, 54

Galapagos Islands 218
Galileo Galilei 24–5, 29, 81,
    231
generation (reproduction)
    208–9, 212, 216,
    217
Genesis
  adherence to 114–15
  Burnet 146, 147, 148, 149,
    151, 152
  creation date 38–9
  Flood 190–1, 195
  geological ideas 141
  Gosse 223, 229
  Hutton 182
  Neptunism 187
  Ray 170, 171
  Whiston 159, 160, 162
  Woodward 154
geological maps 136

Index

geology 108–10, 113–14, 135, 139, 166
Gingerich, Owen 265
glacial theory 193–4
God
  Bible 8
  case for 19, 57
  existence of 8–9, 14, 50
  nature of 233–4, 236–7
  rival views of 40–1
  see also First Cause; Genesis; natural theology
Godwin, William 245, 246–7
Gosse, Edmund 226
Gosse, Philip Henry 223, 224–6, 228–30
  Omphalos 229
Gould, John 218
Grant, Robert 198
gravity 27–8

Harvey, William 12–13
Henslow, John Stevens 2, 4, 140
Henslow, Robert Stevens 271
Herschel, John 5
Hewitt, Jane 15
hierarchical systems *see* classifications
Holmes, Arthur 190
Hooke, Robert
  Boyle 32, 34, 35
  career 33–5
  earthquakes 179, 188
  evolution 197–8

fossils 126–7, 128
geology 113, 163
Micrographia 81, 126
microscopy 81, 82
motion 27
orbital mechanics 178–9
Plot 129
Wren 35
Hooker, Joseph D. 227, 271, 275–6, 277
human condition 238
human eye 80
humans, imperfections 95–6
Humboldt, Alexander von 5
Hume, David
  career 47–51
  cause 18, 102
  Dialogues Concerning Natural Religion 49–50, 59–60, 90
  generation 208
  Malthus 246
  matter, organisation of 205
  nature's imperfection 92, 238
Hutton, James
  career 176–7
  earth history 178, 180–6
  evolution 222
  mountains 187
  Theory of the Earth 38, 39, 183
Huxley, Thomas Henry 224, 227, 270–1, 274–5, 277
hyaenas 191

hybridisation 73
hydra 106–7

ice 193–4
immutability of species 72
imperfection in nature 92, 94–6
inner heat, earth 156, 158, 163,
        168, 187, 189, 190
internal mould 206–7, 208
irreducible complexity 263–5

Jameson, Robert 198
Jefferson, Thomas 135
Jesus 8, 235
Johnson, Dr 45–6
Judeo-Christian tradition 8

Kepler, Johannes 24
Kingsley, Charles 104
Kirkland Cave 193

La Mettrie, Julien Offray de
        104, 105
Lamarck, Jean Baptiste de 198,
        213–15
lathes and files metaphor 100,
        103–5, 106
lava flows 186
Law, Edmund 17
Law, John 17
leClerc, Georges Louis *see*
        Buffon, Compte du

Leewuenhoeck, Antoni 81
Leonardo da Vinci 12, 126
Lhwyd, Edward 130–1
life 97–100, 178, 202, 217
Lightfoot, Sir John 115
Linnaeus 73
Lister, Martin 126, 129, 130,
        205
Locke, John 45
Lubbock, Sir John 271, 275
Lunar Society 36
Lyell, Charles 139, 186, 194

Macfie, Ronald Campbell 262
machine analogy 12
Malthus, Daniel 245–6
Malthus, Thomas Robert 246,
        247, 257
    *Essay on the Principle of
        Population, An* 251–4
maps, geological 136
Marx, Karl 255
mastodons 135
materialist philosophers 205
matter 199
Matthew, Patrick 221
Medici, Ferdinand II de' 164
Mediterranean 195
metaphysics 40, 45
Micrographia (Hooke) 81,
        126
microscopes 81–2
mid-Atlantic Ridge 190
migration, birds 85–6
mind and soul 105

Index

miracles 8, 16, 46–7, 48, 54
moon 25
moral philosophy 40
motion 27, 199
mould, internal 206–7, 208
mountains
  Alps 146
  Burnet 146, 150
  formation 187–9
  Hooke 179
  Hutton 181
  Woodward 154
  mutation 219

natural history 62–3, 68–9,
  85–6
  Oxfordshire 122
natural philosophy 22, 40, 43
  see also science
natural science 73
natural selection 4, 217, 218,
  220–2, 260–1
natural theology
  evolution 216–17
  eye 80
  life 99
  mid-nineteenth century 223
  nature 62–4
  post-Darwinian 261, 263
  premises 6
  Ray 73–5, 76
  science 279
  validity 92
Natural Theology (Paley) 5, 6
  chance 242–3

criticisms, anticipated 13–14
  Darwin, Erasmus 212–13
  economy of nature 241
  generation 208–9
  God, nature of 233–4,
    236–7
  happiness 240–1
  Hume refuted 91, 208–9
  intentions 17, 195–6
  lathes and files metaphor
    100, 103–5, 106
  Malthus 256, 257, 259, 260
  physiology 78
  reactions to 60–1
  success 57
  watch analogy 9–11,
    99–100, 102, 112
nature's imperfection 92, 94–6
Neptunism 187
New Theory of the Earth, A
  (Whiston) 159–62
Newton, Isaac
  Burnet 152
  career 25–7
  laws of motion 27–8
  orbital mechanics 179
  Principia Mathematica 18
  Royal Society 36
Nieuwentyt, Bernard 61–2
Noah's Flood *see* Flood
non-conformists 52–3

Omphalos (Gosse) 229
On the Origin of Species
  (Darwin) 270

orogeny *see* mountains, formation
Owen, Richard 134, 270, 271
Oxford University, science 267–9
Oxford University Museum 268–9, 278, 279

Paine, Thomas 51–2
palaeontology 135
  see also fossils
Paley, William
  Buffon 207, 215
  Cambridge 15
  career 14–18
  Enlightenment 17
  evolution rejected 199, 204
  Final Cause 107–8
  generalisations 87–9
  God, case for 19, 57
  lathes and files metaphor 100, 103–5, 106
  logic 64
  Natural Theology see Natural Theology (Paley)
  Principles of Moral and Political Philosophy, The 15
  View of the Evidences of Christianity, A 15–16, 17
  watch analogy 9–11, 61, 97, 99–100, 102, 112
Paley, William and Elizabeth Clapham 15

Pasteur, Louis 98–9
Pearson, John 2
Pennant, Thomas 85
petrifying springs 123
physico-theology 143
planets, movement of 24
plastick vertue 127, 208, 215
plate tectonics 189–90
Playfair, John 186
Plot, Robert 119–25, 128–30, 132–4
Plymouth Brethren 226
Popper, Karl 234
population 219, 248–50, 251–4
poverty 254
Priestley, Joseph 209–10
Principia Mathematica (Newton) 18
Principles of Moral and Political Philosophy, The (Paley) 15
Principles of Philosophy (Descartes) 143
Prodromus (Steno) 164–8
purpose 42–3

rain 157
Ranelagh, Lady 33
Ray, Elizabeth 66
Ray, John
  career 66–8
  classification 70–1
  faith 73–4

Index

fossils 130–1
  natural theology 64
  Three Physico-Theological
    Discourses 169–73
  Wisdom of God Manifested
    in the Works of the
    Creation, The 73–4, 76,
    77
Redi, Francesco 99
reduction 264
relation 88–9
religion 22, 51–3
religious orthodoxy 141
reproduction (generation)
  208–9, 212, 216, 217
retina 93–4
revelation 51
rocks 109
Rousseau 246
Royal Society 35
Royal Society of Edinburgh
  180
Ruskin, 268, 278

Scheuchzer, Jacob 119
science 22–3, 43–4, 234
  Oxford University 267–9
sea levels 194–5
Second Causes 40, 41, 199,
  242, 263
sects, religious 52–3
Sedgwick, Adam 139, 153
Smith, William 136
Spallanzani, Lazaro 99
species 71–2

immutability of 72
spontaneous generation 98,
  203, 213
springs, petrifying 123
Steno 113, 169
  Prodromus 164–8
Stenson, Niels *see* Steno
Stephen, Leslie 60, 61
stratigraphy 136
sun 23–4, 25
superposition 113
symmetry 87–8

tectonics *see* plate tectonics
telescopes 25, 81, 231
Telluria Theoria Sacra (Burnet)
  62, 142–51, 248
Temple of Serapis, Puzuoli
  139
theism 51
Theory of the Earth (Hutton)
  38, 39, 183
thinking, freedom for 47,
  54
Thirty-nine Articles 52, 66
Three Physico-Theological
  Discourses (Ray)
  169–73
time 114, 184
Toland, John 53–4, 179–80,
  199
Tradescant, John, the Elder
  121
Tradescant, John, the Younger
  121, 122

transmutation 7, 197, 216
  see also evolution
Trembley, Abraham 106–7

Ussher, James 115
utilitarianism 78–9

variation 218–19
*View of the Evidences of
  Christianity, A* (Paley)
  15–16, 17
Vulcanism 187, 188

Wallace, Alfred Russell 221
watch analogy 9–11, 61, 97,
  99–100, 102, 112
  see also lathes and files
  metaphor
watches 11
water, circulation 156

Wedgwood , Emma 4
Wegener, Alfred 189–90
Werner, Abraham Gottlieb
  187
Whiston, William 158–9,
  248–9, 250–1
  New Theory of the Earth, A
  159–62
White, Gilbert 63, 84–7
Wilberforce, Samuel 269, 270,
  271, 272–3, 277
Willoughby, Francis 67–8
*Wisdom of God Manifested in
  the Works of the Creation,
  The* (Ray) 73–4, 76, 77
Woodward, John 153
  *Essay towards a Natural
  History of the Earth*
  153–8
Wren, Christopher 34, 35

Xenophanes 117